Praise for Kenneth C. Davis

Don't Know Much About® Mythology

"Ken Davis is the high school teacher we all wish we'd had—smart, funny, and irreverent. *Don't Know Much About® Mythology* is a crystallized reminder of what's enduring about the past, and why it continues to matter today. It's a perfect companion to Harry Potter, *The Da Vinci Code*, and the Bible—and the best excuse I know to get deserted on a Greek island. Between Odysseus and Icarus, it might even give you a few clues for how to find your way home—and how not to."

—Bruce Feiler, author of *Walking the Bible*

"Because Davis ranges widely and with such sparkling wit through a broad sweep of myths, his survey provides a superb starting point for entering the world of mythology."

—*Publishers Weekly* (starred review)

"An engaging handbook on gods, goddesses, and the civilizations they have inspired. . . . [Davis's] goal as an author is to infect readers with his own intellectual eagerness, and he succeeds admirably with this idiosyncratic tour of world mythology. . . . Even professors will have to concede that Davis has done his research—his annotated bibliography is excellent—and that he's a laudably conscientious scholar. An accessible and informed guide to an always-fascinating subject, and an ideal reference for the general reader."

—*Kirkus Reviews*

"A massive overview of every myth under the sun. Davis shatters commonly held myths about myths, differentiating them from allegories and legends, and explores the history of such tales in societies and religions around the globe, from Mesopotamia's Gilgamesh to Genesis' Noah. You can read here, too, about Native Americans' use of peyote, a tempestuous Nordic god of thunder, and a debate over the meaning of evil."

—*Daily News*

"Who are we? In his thoughtful and entertaining *Don't Know Much About® Mythology*, Kenneth C. Davis suggests that in large part, we are a product of our own creation, our best instincts, and worst prejudices—reflected in the stories we tell. We have become our myths, Davis suggests, though they are not necessarily true. In the Americas, myths justified slavery and the destruction of native societies. Yet myth can empower, pulling us upward toward greater creativity and humanity. For all who choose to know just who we are, you must read this book."

—Richard M. Cohen, author of
Blindsided: Lifting a Life Above Illness

"Over the long development of human culture, the stories of mythology are like a chronology of human evolution. They tell us who we are, and hint at the answer to the growing spiritual intolerance we see today: at the level of the soul, we all want the same things. In *Don't Know Much About® Mythology*, Kenneth C. Davis illuminates these ideas in a popular and entertaining way. I highly recommend this book."

—James Redfield, author of *The Celestine Prophecy*

"With his trademark wit and fiercely entertaining style, Kenneth Davis draws us into mythological worlds, preserving ancient mysteries and enchantments even as he clarifies, orders, and makes sure we have the stories straight. *Don't Know Much About*® *Mythology* frames questions that arouse curiosity and produces answers that lead to astonishment. Whether you want a crash course on North American native myths or a refresher course on Gilgamesh, this book will provide a great read and remain a permanent reference manual."

—DR. MARIA TATAR, Department of Folklore and Mythology, Harvard University, and author of *The Annotated Brothers Grimm*

"In each of his *Don't Know Much About*® books, Kenneth C. Davis has brought the forgotten child to the front row, reminding those of us who hated school that one size doesn't fit all in education—that the desire to learn is far better served by the pursuit of individual passion than by classroom conformity. In *Don't Know Much About*® *Mythology*, Davis uses the intense passion that stirred in his own soul as a fifth-grade boy reading the *Odyssey*, to take us to a place of magic, imagination, and transcendence. Davis not only presents an entertaining exploration of humanity's most sacred stories across many civilizations, he brings us face to face with our most distant ancestors who were driven by innate curiosity to explain life's mysteries. Davis's book is a masterpiece. I couldn't stop turning the pages. "

—ALBERT CLAYTON GAULDEN, founder and director of the Sedona Intensive and author of *Signs and Wonders*

"Davis writes with humor, he can turn a fine phrase. . . . If history were usually taught this way, we wouldn't have to worry about the closing of the American mind."

—*Booklist*

"If you've always wondered exactly what Boss Tweed bossed and what Tammany Hall was, Davis is your man."

—*Washington Post Book World*

Don't Know Much About® Geography

"Davis pulls the reader west and east across the page with humor and a fresh viewpoint."

—*USA Today*

"Makes geography buffs of us all."

—*San Francisco Chronicle*

"Playful . . . fun. . . . His books will go a long way in helping people to answer that simple, important question: 'Where am I?'"

—*Chicago Tribune*

"Davis manages to make learning about the world fun. . . . By the time you finish, you will know more about geography than you ever learned in school."

—*Houston Post*

Don't Know Much About® the Universe

"Beware: You may easily become addicted to Davis's breezy style and his treasury of odd and fascinating facts."
— *Washington Post*

"Covers a huge topic in a lighthearted, conversational question-and-answer format."
— *Atlanta Journal-Constitution*

"A handy reference book, with a slew of good information and a host of age-old questions, intelligently arranged and amusingly addressed."
— *St. Louis Post-Dispatch*

Don't Know Much About® the Civil War

"Highly informative and entertaining. . . . Propels the reader light-years beyond dull textbooks and *Gone With the Wind*."
— *San Francisco Chronicle*

"Lively and relevant."
— *USA Today*

"Distinct, clear, and balanced. . . . Davis has a gift for deftly rendering the essentials."
— *New York Times Book Review*

Don't Know Much About® the Bible

"Do you still postpone reading the Bible cover to cover? *Don't Know Much About® the Bible* offers a rousing companion volume to get you going. Kenneth Davis will take you on a grand tour—with commentary."
— *Christian Science Monitor*

"A great starting point for Bible beginners."
— *San Francisco Chronicle*

"Reading him is like returning to the classroom of the best teacher you ever had!"
— *People*

"Witty and thoughtful, with plenty of surprising yet credible breaths of fresh air on an old subject. Each chapter of Davis's book reads like a letter to an intelligent friend. Pick this one up—you won't be disappointed."
— CHARLES PELLEGRINO, author of *Her Name, Titanic* and *Return to Sodom and Gomorrah*

"One of the most lucid, important, and worthwhile books of this or any year. Here is a perfect book for the dawn of the third millennium. Historian and Renaissance man Kenneth Davis is the perfect writer to tackle this lofty and compelling topic, which he does with zest and gusto, style and spirit, heart, humor, and integrity."
— DAN MILLMAN, author of *Way of the Peaceful Warrior* and *Everyday Enlightenment*

"*Don't Know Much About*® *the Bible* is like a journey through biblical times with a modern tour guide who is fluent in history, psychology, religion, sociology, and anthropology! It's fascinating how Davis gives this ancient text such a rich development. I think his book would be wonderful to read to children."

— CAROL ADRIENNE, author of *The Purpose of Your Life*

"*Don't Know Much About*® *the Bible* is a wake-up call for all of us who slept through Sunday school. Like every great teacher, Ken Davis makes learning an adventure. This ain't mama's Bible filled with begets and begats — it's colorful, entertaining, and filled with incredible facts. Do yourself a favor — read this book."

— MARY MATALIN, author of *All's Fair: Love, War, and Running for President*

© Josef Astor

About the Author

KENNETH C. DAVIS, the *New York Times* bestselling author of *Don't Know Much About® History*, was recently dubbed the "King of Knowing" by Amazon.com. He often appears on national television and radio, and has served as a commentator on NPR's *All Things Considered*. In addition to his adult titles, he writes the Don't Know Much About® Children's series, published by HarperCollins. He and his wife live in New York City and Vermont. They have two grown children. Davis is at work on a new book about American History.

DON'T KNOW
KNOW
MUCH
ABOUT®
anything

THE
DON'T KNOW MUCH ABOUT®
SERIES

by Kenneth C. Davis

BOOKS FOR ADULTS

Don't Know Much About History
Don't Know Much About Geography
Don't Know Much About the Civil War
Don't Know Much About the Bible
Don't Know Much About the Universe
Don't Know Much About Mythology

BOOKS FOR CHILDREN

Don't Know Much About the 50 States
Don't Know Much About the Solar System
Don't Know Much About the Presidents
Don't Know Much About the Kings and Queens of England
Don't Know Much About the Pilgrims
Don't Know Much About the Pioneers
Don't Know Much About Dinosaurs
Don't Know Much About Mummies
Don't Know Much About Planet Earth
Don't Know Much About Space
Don't Know Much About American History
Don't Know Much About World Myths
Don't Know Much About George Washington
Don't Know Much About Sitting Bull
Don't Know Much About Abraham Lincoln
Don't Know Much About Rosa Parks
Don't Know Much About Thomas Jefferson
Don't Know Much About Martin Luther King, Jr.

EVERYTHING YOU NEED TO KNOW
BUT NEVER LEARNED ABOUT PEOPLE,
PLACES, EVENTS, AND MORE!

DON'T
KNOW
MUCH
ABOUT®
anything

KENNETH C. DAVIS

HARPER

NEW YORK • LONDON • TORONTO • SYDNEY

HARPER

Don't Know Much About® is a registered trademark of
Kenneth C. Davis.

DON'T KNOW MUCH ABOUT ANYTHING. Copyright © 2007 by
Kenneth C. Davis. All rights reserved. Printed in the United
States of America. No part of this book may be used or reproduced in any manner whatsoever without written permission
except in the case of brief quotations embodied in critical
articles and reviews. For information address HarperCollins
Publishers, 10 East 53rd Street, New York, NY 10022.

HarperCollins books may be purchased for educational,
business, or sales promotional use. For information please
write: Special Markets Department, HarperCollins Publishers, 10 East 53rd Street, New York, NY 10022.

FIRST EDITION

Designed by Nancy B. Field

Library of Congress Cataloging-in-Publication Data
   Kenneth C. Davis.
   Don't know much about anything / Kenneth C. Davis.
  —1st ed.
     p. cm.

Includes index.

ISBN: 978-0-06-125146-7
ISBN-10: 0-06-125146-1

08 09 10 11 ❖ / RRD 10 9 8 7 6 5

*This books is dedicated to
every child who ever asked "Why."
And to every parent, teacher, librarian, friend, and relative
who might take the time to answer.*

CONTENTS

INTRODUCTION 1

Famous People 5

Exceptional Places 49

Historic Happenings 75

Holidays and Traditions 109

Everyday Objects and Remarkable Inventions 139

Space and the Natural World 161

Sports 183

Entertainment 205

Food 233

Civics 257

AFTERWORD 291

INDEX OF SUBJECTS 293

INTRODUCTION

"Pop quiz, hotshot."

Remember when crazed bomber Dennis Hopper tossed that line at Keanu Reeves in the great thriller *Speed?* Well, now it's your turn.

Actually, those words take most of us back to our school days. And they may still turn your blood cold and make you weak in the knees. Whenever my teachers said, "Okay, class, clear off your desks. Time for a quiz," beads of sweat formed on my brow. It was a moment of pure dread—unless it was a spelling quiz. For some reason, I was always a good speller.

But the truth is—and this may be a bit of a jolt coming from the author of more than twenty *Don't Know Much About* books for adults and kids—I was not a great student. When I was a little boy, no matter how hard I tried, I couldn't sit still in school. Each day, from the get-go of opening bell, I fidgeted and squirmed at my desk, watching the clock and waiting for three o'clock to roll around so that I could hit the playground and ride my bike. I hated textbooks, fought with fractions, and got stomachaches at the simple thought of taking a test.

But here's the odd part—I liked learning and knowing "stuff." I was extremely curious. I loved doing crossword puzzles and word games, like the "Jumbles" in the newspaper, as well as watching game shows that let me test my knowledge. The original daytime *Jeopardy!*, with Art Fleming and Don Pardo, was a "sick day" favorite.

And I loved to read. My bed was always full of books—from the Golden Books I had as a small child to a series of biographies of famous people as children I discovered and read over and over again. I remember getting an encyclopedia on the "installment plan," picking up a new section each week where it was sold in the local supermarket, and putting it into an enormous ring binder.

The trip to the public library in town was practically a ritual each week, and I remember the awe I felt when I moved from the children's room on the ground floor to the majestic marbled adult room upstairs. Whether I was at home or away at summer camp, I ate up comic strips and comic books, everything from Batman and Dare Devil to the Green Lantern and the Fantastic Four. I was also a devoted cereal box reader and always imagined that "Battle Creek, Michigan" must be such an interesting place. And I loved to spread out a map on my lap in the backseat during family trips to places like Gettysburg and Fort Ticonderoga, where I came to see that history wasn't about memorizing dates and speeches, but about real people doing real things in real places.

In other words, learning wasn't what I hated—just the boring way it was usually dished out in school.

Maybe that's why I grew up wanting to make learning fun. When I began to write, I set out to create the kinds of books I wanted to read, and my *Don't Know Much About* series for adults was born out of my personal curiosity and a passion for American history. From there, I moved on to other subjects that have fascinated me since I was young—geography, religion, astronomy, and mythology. Rejecting the widely held notion that these are the "boring" requirements of high school, I always attempted to make these subjects relevant by connecting them to everyday life. Eventually I branched out into books for children and wrote about the presidents, fifty states, the solar system, mummies, world myths, and significant people, among other topics.

But my goal was always the same. And it was simple—to make sure no one's eyes glazed over when they turned the pages. My approach was to ask offbeat questions, bust myths, and bring to light little-known facts that made readers say, "I never heard that. Why didn't they tell us that in school!"

A few years ago, when I became a contributing editor of *USA Weekend*, I was offered a dream assignment—to construct weekly quizzes on lots of different topics. For me, this was an invitation to conspire with readers like myself who groaned when the teacher said, "Okay, class. Put down your pencils. Time's up."

These quizzes cover a wide gamut of subjects, but I don't think they are "trivia." I'd like to believe that anything that is worth knowing is not trivial. So to enjoy this book, all you need is curiosity. No thinking caps, stopwatches, or #2 pencils allowed. And no final grades either. My hope is to get you to agree with the poet William Butler Yeats who said, "Education is not the filling of a pail, but the lighting of a fire."

If you think you don't know much about anything, but you'd like to learn, then welcome aboard.

You've come to the right place, Hotshot.

FAMOUS PEOPLE

DON'T KNOW MUCH ABOUT
Benjamin Franklin

HAD HE ONLY INVENTED bifocals and the stove bearing his name, he would have been notable. If he had only experimented with electricity and charted the Gulf Stream, he would have been a giant of science. If he had only helped draft the Declaration of Independence and the Constitution, he would have been a legend. But Benjamin Franklin did all of these things—and much more. America's first true international celebrity, Benjamin Franklin was born in Boston on January 17, 1706, the fifteenth child in a family of seventeen children, the son of a soap and candle maker. In a remarkable life, Franklin became wealthy, famous, and one of the most important Founding Fathers. When he died at the age of eighty-four on April 17, 1790, nearly twenty thousand admirers attended his funeral. What else do you know about this unique man who helped "invent" America?

1. Franklin was the only person to sign the four key documents that created America. What are they?

2. Which office did Franklin's illegitimate son hold?

3. What did Franklin produce every year for twenty-five years?

4. What did his famous kite experiment prove?

5. What was his greatest accomplishment during the Revolutionary War?

6. Franklin preferred what animal as America's symbol?

7. What was Franklin's final public role?

ANSWERS

1. The Declaration of Independence, the Treaty of Alliance with France, the Treaty of Peace with Great Britain, and the Constitution of the United States.

2. Loyal to England, William Franklin became the Royal Governor of New Jersey. During the Revolutionary War, he was arrested and later went to London.

3. He wrote and published *Poor Richard's Almanac* from 1733 to 1758. Its fame rests on the wit and wisdom that Franklin scattered through each issue.

4. In 1752, he flew a homemade kite during a thunderstorm accompanied by his son William. Franklin proved that lightning is electricity. Then he invented the lightning rod.

5. As a commissioner sent to represent the United States in France, Franklin got the French to join the war against England. Their aid was crucial to America winning its independence.

6. In what may have been his only bad idea, he preferred the turkey to the eagle, which he thought was a bird of bad moral character.

7. In 1787, he was elected president of America's first antislavery society, and his last public act was to sign an appeal to Congress calling for abolition.

DON'T KNOW MUCH ABOUT
Walt Whitman

HE "HEARD AMERICA SINGING." And his work has inspired some, bedeviled others (mostly students), and stood as the work of a unique American voice for more than 150 years. In July 1855, a former schoolteacher turned newspaper publisher, Walt Whitman self-published 795 copies of the first edition of twelve of his poems in a book called *Leaves of Grass.* Over the years, Whitman (1819–1892) would add many more poems to later editions of the work that may be the most famous American book of poems ever published. Sample a bit of Whitman in this quick quiz.

1. Where did Whitman attend college?

2. How did Whitman serve during the Civil War?

3. What event inspired the poems "O Captain! My Captain!" and "When Lilacs Last in the Dooryard Bloom'd"?

4. Why did he lose his government job in 1865?

Answers

1. He didn't. Born on Long Island, New York, he went to school for about six years before becoming a printer's apprentice and was largely self-educated after that.

2. After his brother was wounded in battle, he became a volunteer nurse, aiding the sick in Washington, D.C., hospitals while working for the Army's paymaster's office.

3. The death of Abraham Lincoln, whom Whitman greatly admired.

4. He was fired for his poems, which the Secretary of the Interior found offensive, presumably for some of their homosexual themes.

DON'T KNOW MUCH ABOUT
Rosa Parks

HISTORY IS TAUGHT as the record of presidents, kings, and generals. But sometimes it is the extraordinary story of an "ordinary" person that history must tell. On December 1, 1955, one woman's act of defiance changed history. But it wouldn't be fair to call Rosa Parks, who was born in 1913 in Tuskegee, Alabama, and died October 24, 2005 at age ninety-two, an ordinary person. What do you know about this courageous woman who helped spark the civil rights movement that transformed America? Get aboard this quick quiz.

1. Where and why was Rosa Parks arrested?

2. Before her arrest, was Rosa Parks involved in the civil rights movement?

3. How much education did Rosa Parks, the descendant of slaves, receive?

4. What action did her arrest trigger?

5. Who was elected president of the organization that ran the boycott?

ANSWERS

1. She refused to give up her seat to a white passenger on a Montgomery, Alabama, bus. A city law required that whites and blacks sit in separate rows. The law also required blacks to leave their seats to make room for white passengers.

2. Yes. Rosa Parks had become one of the first women to join the Montgomery chapter of the National Association for the Advancement of Colored People (NAACP) in 1943, serving as its secretary until 1956. Employed as a seamstress, she lost her job as a result of the boycott and later moved to Detroit.

3. She attended Alabama State Teachers College.

4. Her arrest triggered a boycott of the city's segregated bus system that had been planned by local civil rights leaders who were awaiting the right moment. The arrest of Rosa Parks was that moment. For 382 days, thousands of blacks refused to ride Montgomery's buses and the boycott ended when the U.S. Supreme Court declared segregated seating on the city's buses unconstitutional.

5. A young and unknown Martin Luther King, Jr., then a Baptist minister in Montgomery, was chosen as president, providing his first national stage.

DON'T KNOW MUCH ABOUT
Malcolm X

BLACK NATIONALIST LEADER Malcolm X was assassinated on February 21,1965. The focus of Spike Lee's 1992 film starring Denzel Washington, Malcolm X remains one of the most widely admired, yet controversial, African Americans in recent history. *The Autobiography of Malcolm X* was named one of the "Ten Most Important Nonfiction Books of the 20th Century" by *Time* magazine. What do you know about this fiery and charismatic leader?

TRUE OR FALSE?

1. He used "X" as a last name because he was an ex-convict.

2. In 1964, Malcolm X publicly broke with the Nation of Islam.

3. His famed *Autobiography* was written by novelist James Baldwin.

4. While interested in Africa, Malcolm X never traveled there.

5. In 1964, he changed his name again after a pilgrimage to Mecca.

ANSWERS

1. False. He chose "X" as a way to renounce what he considered a "slave name." Born Malcolm Little (1925), he changed his name in 1952 after joining the Nation of Islam, which he learned about while in prison.

2. True. After helping enlarge the Nation of Islam's membership as a highly effective spokesman, Malcolm X left the group in a dispute over leader Elijah Muhammad's extramarital affairs.

3. False. It was a collaboration with journalist Alex Haley, later famous as the author of *Roots*.

4. False. He went to Africa four times. After his fourth trip, which included a pilgrimage to Mecca in Saudi Arabia, he returned to the United States to start the Organization of Afro-American Unity.

5. True. After a conversion to Orthodox Islam, he chose the name El-Hajj Malik. He also came to believe that interracial brotherhood was possible based on his pilgrimage experience.

DON'T KNOW MUCH ABOUT
Houdini

IN MACHPELAH CEMETERY in Ridgewood, Queens, New York, an annual Halloween vigil traditionally marked the death of America's most famous magician and escape artist, Harry Houdini, who died on Halloween in 1926. The legend that Houdini will communicate on the anniversary of his death lives on. There are many such legends about the man who was born Ehrich Weiss in Budapest, Hungary, in 1874. Some of those Houdini myths came from *Houdini*, a 1953 film about his life starring Tony Curtis. What do you know about America's most famous escape artist? Wiggle out of this quiz.

TRUE OR FALSE?

1. The son of a rabbi, Houdini planned to follow in his father's footsteps.

2. His most famous trick was called the Chinese Water Torture.

3. Houdini died during a performance.

4. Houdini was an avid believer in spiritualism.

5. Houdini and his wife shared a secret coded message that he would try to send her if he died.

6. In 2007 a relative of Houdini's decided to try to exhume his body.

ANSWERS

1. False. Although his father was a rabbi in Hungary, young Ehrich Weiss had no such plans. He began doing magic tricks in a "dime museum," but gained international fame doing escapes. The name *Houdini* was a tribute to a famed French magician, Jean Robert-Houdin.

2. True. Houdini could quickly free himself from apparently escape-proof devices, including leg irons, straitjackets, and milk cans. But his most sensational feat consisted of escaping from an airtight tank filled with water. He had the ability to dislocate both shoulders to free himself.

3. False. He allegedly died of complications from a burst appendix, not from the Water Torture as the film depicts it.

4. False. Houdini also became known for criticizing spiritualists who claimed they could communicate with spirits of the dead. He publicly exposed mediums as frauds, causing a rift with his good friend Arthur Conan Doyle, creator of Sherlock Holmes.

5. True. The message was his way to prove that seances and mediums were frauds. His wife continued to hold seances on Halloween for years after his death, beginning a tradition that still goes on. While these seances are still held every year on Halloween, the famed graveside vigil continued until 2005 when the Machpelah Cemetery director prohibited it because it attracted such large crowds. That year, the date of the vigil was changed to November 16, the date of Houdini's death on the Jewish calendar.

6. True. A theory that Houdini may have died of foul play resulted in a court action in 2007 to have his body exhumed.

Mother Teresa

SAY THE NAME *Agnes Gonxha Bojaxhiu,* and you may get blank stares. But *Mother Teresa* (1910–1997) gets instant recognition. This Roman Catholic nun, known as the "saint of the gutters," received the 1979 Nobel Peace Prize for her work with the poor of Calcutta. What do you know about this woman who has inspired legions of devoted followers, along with considerable criticism and controversy?

1. Where was Mother Teresa from?

2. What order did she found?

3. In 2003, what church honor did she receive?

4. What honor did she receive from *Time* magazine?

5. Why do Noble Prize winners receive their awards on December 10?

6. What did Alfred Nobel, founder of the Nobel Prizes, invent?

ANSWERS

1. She was born in what is now Skopje, Macedonia, and went to India in 1931 where she made her vows, took the name Sister Mary Teresa, and later became an Indian citizen.

2. In 1950 she was given permission to start the Missionaries of Charity, whose goal was to care for the hungry, naked, and homeless, those "shunned by everyone."

3. Mother Teresa was "beatified" in 2003, considered a step toward being declared a saint of the Roman Catholic Church, although the required miracle attributed to her has caused considerable controversy.

4. She was named one of the "100 Most Important People of the 20th Century."

5. It is the anniversary of the death of Alfred Nobel, who established the prizes in 1901.

6. The Swiss chemist invented dynamite, patented in 1867. It made him very wealthy and he decided to use his money to reward achievements that benefited humanity.

DON'T KNOW MUCH ABOUT
Michelangelo

FEW PIECES OF ART are as instantly recognizable as *David*, Michelangelo's massive masterpiece. The fourteen-foot statue, carved from a block of marble, was unveiled in Florence, Italy, in 1504. One of the most famous people of his time, Michelangelo (1475–1564) was a genius as a sculptor, painter, and architect during the Italian Renaissance. What else do you know about one of the most revered artists of all time?

TRUE OR FALSE?

1. His first important work was *The Pietà*.

2. Michelangelo never signed his work.

3. The completed *David* was not publicly displayed because of its nudity.

4. *David* still stands in Florence's central plaza.

5. It was illegal for artists to study anatomy in Michelangelo's time.

6. He painted his most famous work, the Sistine Chapel, while lying on his back.

ANSWERS

1. True. At twenty-three, Michelangelo carved a version of the dead Christ in the lap of the mourning Virgin Mary. This statue, now in St. Peter's Basilica in Rome, established him as a leading sculptor. Vandalized in 1972, it has been restored.

2. False. After another artist was given credit for *The Pietà*, Michelangelo chiseled his name into it, but regretted doing so and vowed he would never sign another piece of art.

3. False. The Florentines placed this great work of art in their main public square and it became a source of civic pride. Over time, some copies and reproductions have been made with a strategic fig leaf.

4. False. It was moved indoors to protect it in 1873 and a copy took its place in the plaza in 1882.

5. True. The Church prohibited the study of cadavers, but Michelangelo secretly arranged to do so in exchange for a piece of sculpture.

6. False. Michelangelo began the ceiling in 1508 and completed it in 1512, applying the paint to damp plaster and working quickly before the plaster dried, a technique called fresco. He painted the ceiling standing up on scaffolding, not lying on his back, as widely believed.

DON'T KNOW MUCH ABOUT
Young George Washington

YOU CARRY HIM in your pocket every day. But how much do you really know about the man whose stern face graces the dollar bill and quarter? George Washington was one of the most fascinating, if contradictory, men in American history. Born to a tobacco planter on February 22, 1732, Washington seemed an unlikely candidate for national hero as a young man. If cherry trees and axes are all you remember about the first president and his boyhood, then take a whack at this quick quiz.

TRUE OR FALSE?

1. Washington's father died when he was eleven.

2. Like Thomas Jefferson, Washington attended the College of William & Mary.

3. Washington was a math whiz.

4. Washington made up the cherry tree story himself.

5. Washington's false teeth were made of wood.

ANSWERS

1. True. At the time, George inherited a small farm and ten slaves. Three years later, he went to live with his older half brother Lawrence at the Mount Vernon plantation. His relationship with his mother was distant, and as an adult he never introduced her to his wife or invited her to his home.

2. False. Washington did not attend college and had very little formal education.

3. True. He supposedly loved to count windowpanes and could calculate the number of seeds in a bag of clover. His mathematical skill was one of the reasons he became a surveyor in his teens.

4. False. The story of chopping down the tree was invented after Washington's death by Mason Weems, who wrote a fictionalized biography of Washington meant as a moral lesson for children.

5. False. By the time he became president, Washington had only one of his own teeth left. But he had several sets of false ones made from cow's teeth, hippopotamus ivory, and even human teeth.

DON'T KNOW MUCH ABOUT
Dr. Spock

"TRUST YOURSELF. You know more than you think you do." That simple advice started a revolution. It was the opening of Dr. Benjamin Spock's *Common Sense Book of Baby and Child Care,* the book that raised America's baby boomers. Doling out advice on everything from fevers to thumb sucking, Spock was one of the most influential and sometimes controversial figures in twentieth-century America. Born on May 2, 1903, in New Haven, Connecticut, Spock died in 1998, having left his mark on several generations of American children. What else do you know about "America's pediatrician?"

TRUE OR FALSE?

1. After the Bible, Dr. Spock's parenting guide is the best-selling book in history.

2. When first published in 1946, Spock's book cost twenty-five cents.

3. A Yale graduate, Spock was also a world class sprinter.

4. In the 1960s, Spock was jailed for his antiwar views.

5. Before Spock, America's leading child care manual advised, "Never kiss your children."

ANSWERS

1. True. *The Common Sense Book of Baby and Child Care* has sold more than 50 million copies and has been translated into thirty-nine languages. Revised many times since 1946, the title was later shortened to *Baby and Child Care*.

2. True. In those days, all paperback books cost a quarter and its accessibility and low price helped boost sales.

3. False. A Yale man, he did win a gold medal as a member of a rowing crew in the 1924 Olympics.

4. False. Spock opposed United States involvement in the Vietnam War and, in 1968, was convicted on charges of conspiring to counsel young men to avoid the military draft. But he appealed the verdict and in 1969 his conviction was reversed.

5. True. John B. Watson's *Psychological Care of Infant and Child* contained such stern rules and was used widely in American hospitals.

DON'T KNOW MUCH ABOUT
Gandhi

ON JANUARY 30, 1948, the life of one of the world's most peaceful men ended sadly in violence. At the age of seventy-eight, Mohandas Gandhi, one of the foremost political leaders of the twentieth century, was assassinated on that day. Gandhi had led India to freedom from British control in 1947, using unique methods of nonviolence, and he is honored as the father of the country. Albert Einstein said of Gandhi: "Generations to come will scarcely believe that such a one as this walked the earth in flesh and blood." What do you know about this martyr for peace and tolerance?

1. Why was Gandhi also known as Mahatma?

2. Which famous nineteenth-century American writer influenced Gandhi?

3. What did salt have to do with India's independence?

4. What was "homespun"?

5. Who killed Gandhi?

ANSWERS

1. People called Gandhi the Mahatma, meaning Great Soul. He believed truth could be known only through tolerance and concern for others.

2. Gandhi credited Thoreau's famous essay "Civil Disobedience" as a key influence.

3. In 1930, Gandhi led hundreds of followers on a march to the sea, where they made salt from seawater in protest against the British Salt Acts, which made it illegal to possess salt not bought from the government.

4. Around 1920, Gandhi began a program of hand spinning and weaving cotton. He believed this would help make India self-sufficient and challenge the British textile monopoly which took India's raw cotton and sent it back to India as finished clothing.

5. In January 1948, Gandhi began a fast, hoping to end the bloodshed among Hindu, Muslim, and other groups. When the groups pledged to stop fighting, Gandhi broke his fast. But while on his way to a meeting, Gandhi was shot by a Hindu fanatic who opposed Gandhi's belief in tolerance for all religions.

DON'T KNOW MUCH ABOUT
Henry Ford

YOU COULD HAVE any color you wanted—as long as it was black. On June 16, 1903, Henry Ford started the Ford Motor Company. Unlike Thomas Edison, Ford was no genius inventor, but he helped transform America as few people have. Born on a farm, Ford became a machinist at sixteen and later worked at a Detroit electric company. He became interested in automobiles, then a new invention, building his first in 1896. It was his factory innovations that really shaped the future. What do you know about the man who helped create America's love affair with the car? Crank up this quick quiz.

1. Which Ford classic appeared in 1908?

2. What key innovation did Ford introduce?

3. In 1914, what shocking change did Ford make for his workers?

4. For what political views was Ford later criticized?

5. What did Ford and his son, Edsel, set up in 1936?

ANSWERS

1. The Model T, which became Ford's only model. A simple car that large numbers of people could buy, the car changed little over the years, and, from 1914 to 1925, came only in black. By 1927, Ford sold more than 15 million Model T's.

2. To lower the $825 price, which was too high for most Americans, the company created the assembly line method. Conveyor belts brought parts to workers and each worker performed a particular task, such as adding or tightening a part, instead of completing a car. Production costs fell and the price of a Model T dropped to $550 in 1913 and $290 by 1924, putting the automobile within reach of the average family.

3. He raised the daily minimum wage to $5, more than twice what most workers then earned. He also reduced the workday from nine to eight hours and introduced a profit-sharing plan. Ford wanted workers who could afford to buy his cars, but he also opposed unions and fought hard against attempts to organize his workers.

4. An anti-Semite and isolationist, Ford made many statements critical of Jews and opposed American involvement in World War II.

5. The Ford Foundation, one of the world's largest charitable foundations, which gives grants for education, research, and development.

DON'T KNOW MUCH ABOUT
Ralph Waldo Emerson

BORN ON MAY 25, 1803, Ralph Waldo Emerson was a uniquely American essayist, critic, poet, and popular philosopher. One of the most significant writers in American history, his ideas influenced writers who knew him and generations who followed him. Born in Boston, Emerson was the son of a Unitarian minister. In 1829, he was ordained a Unitarian pastor but resigned his pulpit in 1832 to begin his career as a writer and lecturer that made him famous. What else do you know about this original American thinker? Take this quick quiz.

1. "And fired the shot heard round the world" may be Emerson's most famous line of poetry. What event did it commemorate?

2. What famous writer known for civil disobedience once worked as Emerson's handyman?

3. Fill in the blank: In one famous essay, Emerson wrote, "A _____ _____ is the hobgoblin of little minds."

4. In "Self-Reliance," he wrote "Whoso would be a man, must be a _____."

5. A former minister, Emerson originated a religious philosophy. What was it?

ANSWERS

1. "The Concord Hymn: Sung at the Completion of the Battle Monument, July 4, 1837," was composed in 1836, sixty years after the battles of Concord and Lexington that started the American Revolution on April 19, 1775.

2. Henry David Thoreau, who graduated from Harvard and met Emerson, who encouraged him to write, gave him useful criticism, and employed him as a gardener.

3. The missing words are "foolish consistency" from the essay "Self-Reliance."

4. Nonconformist.

5. Transcendentalism. He favored a new religion founded in nature and fulfilled by direct, mystical intuition of God. Transcendentalists believed that organized Christian churches interfered with the relationship between a person and God.

DON'T KNOW MUCH ABOUT
J.R.R. Tolkien

HAVING CAPTIVATED READERS for more than fifty years, the magical work of John Ronald Reuel Tolkien (1892–1973) has come to life for a generation of moviegoers. J.R.R. Tolkien (pronounced TOHL keen) was the English author and scholar who created the mythical world known as Middle-earth. After serving in World War I, Tolkien settled into a long teaching career but devoted his life to his great creation, best known in the form of the three related novels called *The Lord of the Rings*. What do you know about the creator of this epic fantasy of good versus evil?

TRUE OR FALSE?

1. Tolkien was a mathematician and code breaker at Cambridge University.

2. Tolkien's first book about Middle-Earth was called *The Silmarillion*.

3. The short, hairy-footed creatures called hobbits first appeared in *The Hobbit* (1937).

4. Written largely during World War II, *The Lord of the Rings* was not published until after Tolkien's death.

5. Before the recent hit films, there were two other screen adaptations of Tolkien's work.

6. Tolkien was a member of a well-known singing group called The Inklings.

ANSWERS

1. False. Born in South Africa of English parents, Tolkien taught medieval languages and literature at Oxford University in England from 1925 to 1959. His hobbit stories were greatly influenced by medieval English, German, and Scandinavian language and legends.

2. True. In 1917, Tolkien began to write *The Silmarillion*, a history of Middle-Earth before the hobbits appeared, but he died before completing it. His son Christopher completed the novel, published in 1977.

3. True. In *The Hobbit*, Bilbo Baggins, a hobbit, discovers the famous ring that conveys the power of invisibility but also corrupts the user.

4. False. The book was broken into three novels. *The Fellowship of the Ring* and *The Two Towers* appeared in 1954 and *The Return of the King* in 1955.

5. True. Animated versions of *The Hobbit* and the first portion of *The Lord of the Rings* were made in the 1970s.

6. False. The Inklings were a group of Oxford writers specializing in fantasy, which also included C. S. Lewis.

DON'T KNOW MUCH ABOUT
Sitting Bull

ON JULY 20, 1881, soldiers at Fort Buford (in present-day Montana) were surprised by the surrender of 188 starving, poorly clothed Indians. Led by the great chief Sitting Bull, they were the remnants of the much feared Hunkpapa Lakota, or Sioux, who helped defeat Custer at the battle of Little Bighorn on June 25, 1876. Born in what is now Bullhead, South Dakota (1831–1834?), Sitting Bull was called Slow (Hunkesni) as a boy. After showing great bravery in a fight against the rival Crow, he received the name Sitting Bull (Tatanka Iyotanka). Known for his courage as a warrior, Sitting Bull was also a visionary holy man. Sitting Bull was killed by Indian reservation police when they attempted to arrest him in December 1899. What else do you know about this great Native American leader who tried to keep his people free?

1. Did Sitting Bull kill Custer?

2. What happened to Sitting Bull after Little Bighorn?

3. After his surrender, how did Sitting Bull earn a living?

4. Who were the Ghost Dancers?

Answers

1. No. That was the popular myth at the time, but Sitting Bull did not even fight in the battle. Sitting Bull acted only as the leading holy man in the preparations for the battle. The year before, he had received a vision that all his enemies would be delivered into his hands. The Indians were led in battle by Crazy Horse and Chief Gall.

2. After the battle, Sitting Bull and his followers were driven into Canada, which they knew as "Grandmother's Land," for Queen Victoria, where they were permitted to live and hunt buffalo.

3. He briefly farmed and raised cattle, but was more famous when he became the star attraction in Buffalo Bill Cody's Wild West Show, which toured America in 1885.

4. In 1890, Sitting Bull helped revive the Ghost Dance, a tribal religion that proclaimed that the Indians would recover their lost lands. The U.S. government thought this was an attempt to renew the Indian wars, and sent half the army to the reservations. When some Indian reservation police were sent to arrest Sitting Bull, he and his son were shot and killed.

DON'T KNOW MUCH ABOUT
Queen Elizabeth II

WHEN GREAT BRITAIN'S monarch celebrated half a century on the throne in 2002 with a Silver Jubilee Festival, it must have been a bittersweet moment for Queen Elizabeth II, who had recently lost her mother and sister. Americans, who threw off the British monarchy in 1776, still have a special fascination for England's enduring institution. What do you know about the fortieth monarch since William the Conqueror in 1066, to whom she is related? Test your regal wits.

TRUE OR FALSE?

1. Elizabeth's fifty-plus years on the throne are the longest reign for a British Queen.

2. The Queen or any English monarch can still overrule the Parliament.

3. During her reign, there have been eleven different prime ministers.

4. No Pope has visited Britain since the Protestant Reformation.

5. The royal name Windsor was made up in 1917.

ANSWERS

1. False. Queen Victoria ruled for sixty-three years, the longest of any British monarch.

2. False. The United Kingdom is a constitutional monarchy. Queen Elizabeth II is the head of state, but the prime minister is the head of the government and Parliament is the chief law-making body.

3. True. The first was Winston Churchill. Prime Minister Tony Blair, the tenth, was the first born during Elizabeth's reign.

4. False. In 1982, Pope John Paul II became the first to go to Britain in 450 years.

5. True. Windsor, the name of the present royal family, was adopted in 1917, and was taken from Windsor Castle, a royal residence. It was chosen to replace Saxe-Coburg-Gotha, which was abandoned during World War I because of its German origin. In 1960, Queen Elizabeth II announced that future generations, except for princes and princesses, will bear the name Mountbatten-Windsor in honor of her husband, Philip Mountbatten.

DON'T KNOW MUCH ABOUT
Anne Frank

"IN SPITE OF EVERYTHING, I still believe that people are really good at heart." Those famous words, written by a teen-ager hiding for her life, have affected millions of people. In 1952, *The Diary of Anne Frank* was published in America for the first time. (Originally published in 1947, the diary was not translated into English for several years.) A German-Jewish girl, Anne Frank was given a blank diary on her thirteenth birthday and wrote a vivid, tender account of her experiences during two years of hiding from the Nazis during World War II. It contained a message of tolerance, courage, and hope, but was also the personal journal of a remarkably sensitive young writer. What do you know about this young woman who has touched the lives of millions of readers?

1. Where was Anne Frank born?

2. Where did Anne and her family hide?

3. How were they discovered?

4. What happened to Anne Frank?

5. How many copies of the book in how many languages have been sold?

ANSWERS

1. In Frankfurt, Germany, on June 12, 1929. She and her family moved to the Netherlands in 1933 after the Nazis began to persecute Jews.

2. In 1942, during the Nazi occupation of the Netherlands, her family, along with several other people, hid in a secret annex behind the Amsterdam office of her father's business. They were aided by loyal employees of Anne's father.

3. Two years later, the family was betrayed by an anonymous caller and arrested.

4. Anne died of typhus in 1945 in the Bergen-Belsen Nazi concentration camp.

5. Since it was first published in 1947, the book has been translated into more than sixty-five languages and has sold more than 20 million copies worldwide. In 1995, a complete and uncut version was published revealing Anne Frank's more intimate thoughts, which had been edited out of the diary by her father.

DON'T KNOW MUCH ABOUT
Helen Keller

PEOPLE WHO CHANGE history are supposed to be politicians and generals, not little girls. But one child born in Tuscumbia, Alabama, on June 27, 1880, certainly made a difference in the world. After an illness destroyed Helen Keller's sight and hearing as an infant, she lived for the next five years as a kicking, screaming wild child. In 1887, Anne Sullivan (1866–1936), child of poor Irish immigrants and nearly blind herself, was hired to tutor the uncontrollable Helen. Through touch, Sullivan was able to reach Keller. Using a manual alphabet in which words were spelled out in her hand, Keller gradually learned to read and write Braille, eventually learned to speak, and went on to college. As a writer and speaker, she crusaded to improve conditions for the blind and deaf-blind until her death in 1968. What do you know about this heroic conqueror of physical disabilities? Take this quick quiz.

1. What famous American inventor advised Helen's father to seek help at Boston's Perkins Institution for the Blind?

2. What college did Helen Keller attend?

3. How did Helen Keller "listen" to people?

4. In the 1962 film of Helen's story, *The Miracle Worker*, Helen was played by Patty Duke, who won an Oscar, and Anne Bancroft portrayed Anne Sullivan. Who played the Sullivan role in a 1979 television remake?

ANSWERS

1. Alexander Graham Bell, inventor of the telephone, whose wife was also hearing impaired.

2. She went to Radcliffe in Cambridge, Massachusetts, and graduated in 1904 with honors. Sullivan assisted her through her college years, interpreting lectures.

3. She "read" lips by touching the lips and throat of people as they spoke.

4. Patty Duke. The role of Helen was taken by Melissa Gilbert.

DON'T KNOW MUCH ABOUT
Abraham Lincoln

A ROUGH-HEWN CABIN, eighteen feet long, near a creek on Sinking Spring Farm. It had a packed dirt floor, a single window, and a door that swung on leather hinges. It wasn't much of a grand entrance on February 12, 1809 for the man who became America's greatest president. While plenty of politicians have claimed humble beginnings, in Abraham Lincoln's case, it was a harsh reality. At age nine, he lost his mother, one of many tragedies he would face in his life. His father remarried and uprooted the family in search of better opportunity, which never seemed to come. Despite limited schooling, Lincoln fell in love with reading and once said, "My best friend is the man who'll get me a book I ain't read." Be honest—what else do you know about "Abe?" (A nickname of which he was not fond.)

1. What law school did Lincoln attend?

2. What military experience did wartime President Lincoln have?

3. What did the Emancipation Proclamation actually do?

4. How many children did Lincoln have?

ANSWERS

1. None. Lincoln estimated he had about one year of formal schooling. Passing the bar in Illinois only required evidence of moral character and an enrollment by a clerk of the State Supreme Court, although Lincoln was well read in the law.

2. He served briefly as a volunteer in the Black Hawk War of 1832, but saw no action except for "fighting mosquitoes in an onion patch."

3. It did not free all the slaves. It freed only those slaves in Confederate territory not under Union control. Full emancipation came with the passage of the 13th Amendment.

4. Three sons survived infancy. Robert Todd, who became a prominent businessman; Willie, who died at twelve, the only president's child to die in the White House; and Tad, who died at eighteen, five years after his father's assassination.

DON'T KNOW MUCH ABOUT
Amelia Earhart

IN THE LIFE OF an extraordinary woman who made history, January 12 stands as a compelling landmark. On that date in 1935, pioneering flier Amelia Earhart became the first person to fly across the Pacific from Hawaii to California. Ten other pilots had already died attempting the feat. Born in Kansas in 1897, aviatrix and feminist Amelia Earhart was nicknamed "Lady Lindy" by the press for her daring exploits as a pilot in days when women did not do such things. In 1937, at the peak of her fame, she and navigator Fred Noonan disappeared while attempting to fly around the world; their fate is one of the century's great mysteries. What do you know about one of the most extraordinary women of the twentieth century? Take this quick quiz.

TRUE OR FALSE?

1. In 1928, Earhart became the first woman to fly a plane across the Atlantic.

2. In 1929, Earhart organized a race for women pilots called "The Powder-Puff Derby."

3. The remains of Earhart and Noonan were never discovered.

4. It is now known that Earhart was shot down and captured by the Japanese while on a secret spy mission.

ANSWERS

1. False. Her first transatlantic flight in 1928 was as a passenger. She became the first woman to cross the Atlantic solo in 1932, exactly five years after Charles Lindbergh had accomplished that feat.

2. True. It was a cross-country air race, and humorist Will Rogers gave it the name, which stuck.

3. True. Although there have been reports of bones found on a remote island, these rumors were never substantiated. Earhart's final radio report came over the Pacific, after she left New Guinea, halfway through her globe-circling trip. Most historians believe that Earhart and Noonan crashed into the Pacific Ocean and did not survive.

4. False. Although this scenario has been suggested, there is no evidence to support it.

DON'T KNOW MUCH ABOUT
Thomas A. Edison

From rooftop decorations and twinkling lights on trees, to the candles on menorahs and that luminous ball lowered on New Year's Eve at Times Square, December is the season of lights. Electric lights, that is. How appropriate to recall that inventor Thomas Edison first publicly demonstrated the incandescent electric light late in December 1879. (The incandescent lamp had actually been perfected in Edison's labs on October 21, 1879.) Need some more batteries for those toys? Thank Edison. A school dropout with little formal education, Thomas Alva Edison (1847–1931) changed the world. More important than the light itself was Edison's development of large central power stations. With more than one thousand patents to his name—many the work of men in his laboratory—Edison combined the imagination of a genius with a showman's feel for marketing and great instincts for commercial ideas. How bright are you when it comes to the amazing Edison and his inventions?

TRUE OR FALSE?

1. Edison received his first patent in 1868 for a vote counter intended to speed up proceedings in Congress.

2. Edison's first successful invention was the stock ticker.

3. The discovery of the carbon light filament was the result of a lab accident.

4. In December 1877, Edison produced the first phonograph.

Answers

1. True. It was rejected because members of Congress felt that slow votes had some advantages.

2. True. In 1869, Edison sold his stock quote device to a Wall Street firm for $40,000, then a sizable fortune, and used the money to set up the world's first industrial research laboratory.

3. False. Edison and his team had patiently tested six thousand different filament materials before hitting on carbonized thread.

4. True. His phonograph was patented on December 15, 1877. He would later combine this invention with his "kinetoscope" to develop the motion picture projector.

DON'T KNOW MUCH ABOUT
Albert Einstein

"WHEN A MAN SITS with a pretty girl for an hour, it seems like a minute. But let him sit on a hot stove for a minute and it's longer than any hour." That is how Albert Einstein once summed up the concept of relativity. One man who changed the world, Einstein was born on March 14, 1879 in Ulm, Germany. A reluctant student who cut class to study physics on his own, he revolutionized science, for which he won the Nobel Prize for physics in 1921. What else do you know about one of the great minds in history? You don't need to be an Einstein to take this quick quiz.

1. What was Einstein's job when he published his 1905 paper containing the special theory of relativity?

2. What does the "c" in $E=mc^2$ mean?

3. Why did Einstein come to America in 1933?

4. What did a 1939 Einstein letter to President Roosevelt inspire?

5. What dimension did Einstein add to the three dimensions of classical geometry?

ANSWERS

1. He was a clerk in the Swiss patent office where he had worked since 1902. His first academic post did not come until 1909.

2. The velocity of light. The equation, which anticipated the splitting of the atom, means that the energy (E) in matter equals its mass (m) times the square of the velocity of light.

3. Einstein, who was Jewish, left Germany when Hitler came to power.

4. The Manhattan Project, which developed the first atomic bomb.

5. Time, often described in modern physics as "space-time."

EXCEPTIONAL
PLACES

DON'T KNOW MUCH ABOUT
the Panama Canal

"A MAN. A PLAN. A canal. Panama." (Read it backwards!) The canal that cuts across the Isthmus of Panama, linking the Atlantic and Pacific oceans, ranks as one of the world's greatest engineering achievements. On December 31, 1999, Panama took control of the canal, ending nearly a century of controversial United States control. Panama was once part of neighboring Colombia, but, in 1903, some Panamanians revolted and declared their independence, with help from President Theodore Roosevelt and some U.S. Marines. Work on the canal began soon after. Digging through disease-ridden jungles and swamps, thousands of laborers, many of whom died, worked for about ten years. The first ship passed through in 1914, but a grand opening was delayed by World War I. What do you know about this "path between the seas"? Dig into this quick quiz.

1. What is an isthmus?

2. How long was the sea voyage before the canal was completed?

3. How long is the Panama Canal?

4. If you go through the canal from the Atlantic to the Pacific, in what direction are you sailing?

5. Who was William C. Gorgas?

ANSWERS

1. A narrow strip of land connecting two larger masses of land.

2. Previously, ships had to travel around South America—a distance of more than 15,100 miles. The canal shortened a ship's voyage between New York City and San Francisco to less than 6,100 miles.

3. The Panama Canal extends 50.72 miles from Limon Bay on the Atlantic side to the Bay of Panama on the Pacific side.

4. A ship traveling through the canal from the Atlantic to the Pacific sails from northwest to southeast. The ship actually leaves the canal twenty-seven miles east of where it entered.

5. The medical hero of the canal, Colonel William C. Gorgas was an American doctor famous for wiping out yellow fever in Cuba after the Spanish-American War. He was in charge of improving sanitary conditions in the Canal Zone and wiped out yellow fever and eliminated the rats that carried bubonic plague in the Canal Zone. By 1913, he had also reduced the high death rate caused by malaria.

DON'T KNOW MUCH ABOUT
the Everglades

FEELING SWAMPED? On December 6, 1947, President Truman created the Everglades National Park. Made up of vast saw grass prairies, deep mangrove swamps, subtropical jungle, and the warm waters of Florida Bay, the park is located on the southern tip of the Florida peninsula and is the only subtropical preserve in North America. What else do you know about one of America's natural wonders? Take this quick quiz.

TRUE OR FALSE?

1. The Everglades were created by the Army Corps of Engineers.

2. The National Park takes up less than a quarter of the original Everglades area.

3. The Everglades are home to both alligators and crocodiles.

4. For centuries, the area was uninhabited.

5. At the northern border of the Everglades lies one of the country's largest lakes.

Answers

1. False. The Everglades were created about ten thousand years ago, after the last major ice sheet melted, raising the level of the sea, which flooded the outlets of Everglades streams and turned the area into a swamp.

2. True. Everglades National Park covers 1,506,499 acres, which makes up about one-fifth of the Everglades' original area.

3. True. It is the only place in the world where alligators and crocodiles exist side by side. The park is also home to 125 species of fish and more than one thousand species of plants, and is known for its rich bird life, particularly large wading birds. Offshore species include the West Indian manatee and bottlenose dolphins.

4. False. Various peoples have lived in the Everglades through the centuries. The Seminole Indians fled to the area in the early 1800s during a period of wars against United States troops.

5. True. Lake Okeechobee, which covers about seven hundred square miles, is the largest lake in the southern United States. Okeechobee is a Seminole Indian word that means "plenty big water." The older Native American name for the Everglades is Pahayokee, or "grassy waters."

DON'T KNOW MUCH ABOUT
the Rocky Mountains

North America's largest mountain system, the Rocky Mountain Chain extends more than three thousand miles through the United States and Canada, stretching from New Mexico through Colorado, Utah, Wyoming, Idaho, Montana, Washington, and Alaska. In Canada, the Rockies run through Alberta, British Columbia, the Northwest Territories, and the Yukon Territory. What do you know about this spectacular range? Explore this quick quiz.

TRUE OR FALSE?

1. The Rockies form the Continental Divide.

2. Pikes Peak is the tallest peak in the American Rockies.

3. Utah's highest peak is Kings Peak.

4. Most peaks in the Rockies were formed by volcanic activity.

5. The famed pioneer route, the Oregon Trail, detoured around the Rockies.

6. The Eisenhower Memorial Tunnel is the world's highest motor traffic tunnel.

Answers

1. True. The Divide separates rivers that flow west to the Pacific Ocean from those going east to the Atlantic Ocean. The Canadian Rockies also separate rivers flowing north to the Arctic Ocean from those that empty into the Pacific. The Arkansas, Colorado, Columbia, Missouri, and Rio Grande are among the rivers that begin in the Rockies.

2. False. Mount Elbert in Colorado rises 14,433 feet. Colorado has forty-six peaks more than fourteen thousand feet high.

3. True. It is 13,528 feet high.

4. False. They were formed millions of years ago during a great upheaval of the earth's crust, and the sides of the mountains contain fossils of animals that once lived in the sea. The southern half of the Rockies does include mountains that were once volcanic plateaus.

5. False. The Rockies slowed the way west during the 1800s, but the Oregon Trail, the long overland route used by pioneers, wound through the Rockies.

6. True. Located west of Denver, the tunnel has an altitude of about eleven thousand feet.

DON'T KNOW MUCH ABOUT
Vietnam

When Saigon fell to North Vietnamese forces on April 30, 1975, it marked the end of a chapter in a long, tragic era. Decades of war cost more than fifty-eight thousand American and millions of Vietnamese lives. The legacy of America's involvement in Vietnam continues to influence American policy and politics. In 1995, the United States extended full diplomatic recognition to Vietnam. What else do you know about a country that was once on America's nightly news with scenes of war but has attracted attention of late for its economic success.

1. What is Saigon called today?

2. Which city is the capital of Vietnam?

3. What country dominated Vietnam for more than one thousand years?

4. What country controlled Vietnam from the 1860s until 1954?

5. True or False: Since the war ended in 1975, more than forty thousand Vietnamese have been killed by "Unexploded Ordnance."

ANSWERS

1. The country's largest city is now called Ho Chi Minh City, after the revolutionary leader who led North Vietnam.

2. Hanoi, the capital of North Vietnam, is now the nation's capital.

3. China's Han dynasty conquered Vietnam in 111 B.C. and the Chinese ruled the country until A.D. 939 when the Vietnamese gained independence.

4. The French controlled Vietnam as a colony beginning in 1858. The Japanese held it during World War II.

5. True. According to UNICEF and other non-government organizations, land mines and other munitions left over from the war have killed more than forty thousand Vietnamese, many of them children. Another sixty-four thousand have been injured.

DON'T KNOW MUCH ABOUT
London

STRAWBERRIES, stiff upper lips, and some wet British weather typically go with the "fortnight" at Wimbledon, the grand dame of tennis tournaments. First played at the All England Lawn Tennis Croquet Club in 1877, the championship spotlights the much older traditions of London. What do you know about this ancient city, once seat of an empire and still one of the world's financial, political, and cultural centers?

TRUE OR FALSE?

1. London was established by the ancient Celtic priests called Druids.

2. The city claims the world's first subway.

3. The London Eye is England's oldest pub.

4. London Bridge never really "fell down."

5. During World War II, "the Blitz" killed more than twenty thousand Londoners.

ANSWERS

1. False. Although there were Celtic settlements in the area, the Romans founded Londinium on the north bank of the Thames River in A.D. 50.

2. True. London's underground railway is the oldest in the world and one of the world's busiest.

3. False. Known as the Millennium Wheel, the "Eye" is the world's largest observation wheel—not a ferris wheel—and opened to the public in March 2000. It was once voted the world's top tourist attraction in a poll.

4. False. There have been many London bridges going back to Roman times. But the one that inspired the nursery rhyme was probably a wooden structure that was washed away by a flood. In 1176, a stone bridge was built and another London Bridge was constructed in 1825. That's the one that was sold and transported to Lake Havasu City, Arizona, in 1973.

5. True. According to the Museum of London, the aerial bombing of London by the Nazis between September 1940 and May 1941 cost at least that many lives and left 1.5 million people homeless.

DON'T KNOW MUCH ABOUT
Hiroshima

"IT SEEMS TO BE the most terrible thing ever discovered," Harry Truman wrote in his diary. "But it can be made to be the most useful." In 1945, President Harry S. Truman wrote those words as he contemplated the most awesome decision of his new presidency: the use of the atomic bomb against Japan. On August 6, 1945, a United States Army plane named Enola Gay dropped a single atomic bomb nicknamed "Little Boy" on the center of the city. (Three days later, a second larger bomb named "Fat Man" was dropped on the city of Nagasaki.) What do you know about this day in history?

1. Where is Hiroshima?

2. What was the toll in Hiroshima? Nagasaki?

3. When did the war with Japan end?

4. What is the Peace Memorial Park?

5. What is the Atomic Bomb Dome?

ANSWERS

1. The city is on Honshu island and occupies a group of islands formed by a river delta. Originally a fishing village, it became one of Japan's largest cities. By World War II, Hiroshima was a military center, but with a civilian population of about 280,000 people.

2. The first atomic bomb, which destroyed about five square miles of Hiroshima, killed between seventy thousand and one hundred thousand initially; the toll from the effects of atomic radiation reached two hundred thousand according to a post-war survey. Nagasaki, located on the island of Kyushu and the first Japanese port opened to Westerners, was an important shipbuilding center. The blast there killed an estimated forty thousand people initially and more than one hundred forty thousand people died within five years.

3. On August 14, Japan agreed to surrender. The formal terms were signed on September 2, which President Truman declared V-J Day for Victory over Japan Day. The surrender was made aboard the battleship *Missouri* and was accepted by General Douglas MacArthur.

4. Ground Zero of the bomb blast in Hiroshima is now a public park where a memorial service is held each year on August 6.

5. A building that was left unreconstructed after the war, it has become a symbol of the peace movement.

DON'T KNOW MUCH ABOUT
the Great Smoky Mountains

FORMING THE BORDER between North Carolina and Tennessee, the Great Smoky Mountains are among the highest and most ruggedly beautiful mountains in the Appalachians, North America's second longest mountain range. To honor and preserve that beauty, Great Smoky Mountains National Park was created on June 15, 1934. (The park was not officially dedicated until September 2, 1940.) What do you know about this majestic park, which attracts millions of visitors and was named a World Heritage site and a UNESCO International Biosphere Reserve? Remember, only you can prevent wildfires and take this quick quiz.

TRUE OR FALSE?

1. The Great Smoky Mountains got their name from the frequent forest fires there.

2. The 521,000 acre Great Smoky Mountains National Park is the most visited site in the National Parks system.

3. The highest peak in the Smoky Mountains is Clingmans Dome.

4. Cherokee Indians were the region's first settlers.

ANSWERS

1. False. The Great Smoky Mountains are named because they are often covered by haze. The park is thickly forested with more than two hundred species of trees, which create a dense, humid atmosphere that looks like a smoky mist.

2. False. While it is the most visited National Park—with more than 9 million visitors in 2004—Great Smoky Mountains National Park trails both the Blue Ridge Parkway in Virginia and Golden Gate Recreation Area in California (neither of which is a national park) in recreational visits, according to National Park Service figures.

3. True. Clingmans Dome is 6,643 feet tall.

4. True. The Smokies are the ancestral home of the Cherokee. The discovery of gold on Cherokee lands led to their removal and the tragic Trail of Tears. More than fourteen thousand Cherokee were forced west in 1838, and nearly five thousand of them perished on the journey. A few Cherokee refused to move and hid in the mountains. Called the Eastern Band, they claimed some land in North Carolina called the Qualla boundary.

DON'T KNOW MUCH ABOUT
Berlin

WITH THE END of World War II in 1945, the city of Berlin was divided between Communist east and democratic west. For years it served as a tense symbol of the Cold War. Following Germany's reunification in 1990, Berlin was named capital of the reunited Germany, and five years ago the German government returned to Berlin. What do you know about this eight-hundred-year-old city that was once the focus of Cold War divisions but successfully hosted the 2006 World Cup finals?

TRUE OR FALSE?

1. Berlin is Germany's largest city.

2. The notorious Berlin Wall was begun in 1948.

3. During World War II, Berlin suffered severe casualties.

4. After the Berlin Wall was built, more than 170 people died trying to escape East Berlin.

5. In his famous "Tear down that wall" speech, Ronald Reagan also said, "All free men, wherever they may be, are citizens of Berlin."

ANSWERS

1. True. One of Europe's great cultural, political, and economic centers, Berlin is Germany's largest city with a population of more than 3.3 million in 2005. It is followed by Hamburg, Munich (Munchen), Frankfurt, and Cologne (Koln).

2. False. East Germany began building the concrete and barbed wire wall to divide east from west in August 1961.

3. True. Allied bombing raids and heavy fighting nearly wiped out Berlin, and some 152,000 civilians lost their lives.

4. True. Most were shot by East German border guards.

5. False. That was how John F. Kennedy began his famous "Ich bin ein Berliner" quote in June 1963. A *berliner*, by the way, is actually known to Germans as a breakfast pastry, according to a report in the *New York Times* in 1988.

DON'T KNOW MUCH ABOUT
Mexico

ON SEPTEMBER 15 each year, Mexico's president traditionally rings a bell and repeats a famous speech called the "Grito
de Dolores" ("Cry of Dolores"), leading Mexicans to celebrate
September 16 as their Independence Day. A republic for nearly
two centuries, Mexico has many traditions dating back to the
ancient civilizations of the Maya and Aztec. What do you know
about America's southern neighbor?

TRUE OR FALSE?

1. The "Grito de Dolores" called for a rebellion against
   Spain in 1810.

2. Mexico's land area is about one-half of the United
   States.

3. Mexico is the world's fifth most populous country.

4. The Mexican Republic shares borders with the United
   States on the north and Nicaragua and Panama to the
   south.

5. Mexico City is home to the oldest university in North
   America.

6. President Vicente Fox's party has been in power in
   Mexico since 1910.

ANSWERS

1. True. Late on September 15, 1810, Miguel Hidalgo y Costilla, a priest, made a speech in his church in the town of Dolores calling for a rebellion against Spain. Although independence was not achieved until 1823, Father Hidalgo's speech is viewed as the moment of Mexico's declaration of independence from three centuries of Spanish rule.

2. False. With some 741,600 square miles, Mexico is about three times the size of neighboring Texas, but much smaller than America's fifty states.

3. False. With some 106 million people in 2005, according to U.S. Census Bureau rankings, Mexico has the eleventh largest population in the world. The Top Ten populations in 2005 were China, India, the United States, Indonesia, Brazil, Pakistan, Bangladesh, Russia, Nigeria, and Japan.

4. False. Guatemala and Belize are Mexico's southern neighbors.

5. True. The University of Mexico, the largest university in Mexico, was founded in 1551 as the Royal and Pontifical University. It was later taken over by the government but is now independent.

6. False. The elections held in July 2000 marked the first time since the 1910 Mexican Revolution that the Institutional Revolutionary Party (PRI) lost the presidency. Vicente Fox of the National Action Party (PAN) was sworn in as the first chief executive elected in free and fair elections. Fox's term ran through 2006, at which time he was ineligible for reelection under Mexico's constitution. Following a controversial and close election, Fox was succeeded by Felipe Calderón, also of the PAN, who narrowly defeated Lopez Óbrador of the Party of the Democratic Revolution (PRD).

DON'T KNOW MUCH ABOUT
Washington, D.C.

The year 2000 marked the bicentennial of the nation's capital city. Congress first met in the "Federal City" in November 1800, and the District of Columbia was officially designated as the seat of government of the United States on December 12, 1800. The idea for the first planned national capital dates back to 1783, when the Congress, in Philadelphia, decided that a permanent capital was needed and then debated it until the Residence Act was passed in 1790. When the federal government relocated to the ten-mile-square site in 1800, it was a town of some twenty-five hundred residents, more than six hundred of them slaves. Since then, it has inspired mixed feelings. Lafayette once called D.C. "the central star of the constellation," while one wit quipped, "If you want a friend in Washington, buy a dog." What else do you know about the nation's capital?

1. What two rivers bound the city?

2. Who designed the nation's capital?

3. Who was Benjamin Banneker?

4. What other Washington, D.C., institution was established two hundred years ago with a $5,000 appropriation from Congress?

5. What did the landmark Compromise of 1850 ban in Washington, D.C.?

6. What were residents of Washington allowed to do in 1964 that they couldn't do before?

ANSWERS

1. The Potomac and Anacostia rivers.

2. French architect Pierre L'Enfant was selected by George Washington, who had selected the site for the capital, as master planner. Much of his grand scheme was not completed before the end of the century. A major redesign of the city was then carried out in the so-called 1901 Plan, which included the Lincoln Memorial.

3. A free African American from Maryland, Banneker was a mathematician and astronomer who helped Andrew Ellicott survey and lay out the future capital.

4. The Library of Congress, which was burned by the British in 1814. The current library was constructed in 1896.

5. The slave trade, but not slavery itself, which was abolished in the city after January 1851.

6. Vote in presidential elections. Washington, D.C., still has only a nonvoting delegate in the House of Representatives.

DON'T KNOW MUCH ABOUT
Mardi Gras in New Orleans

THE DEVASTATING EFFECTS of Hurricane Katrina in late August 2005 may have changed New Orleans forever. The harsh realities of the recovery there will long challenge a city whose unofficial motto has always been *Laissez le bon temps roulez.* ("Let the good times roll.") New Orleans is home to jazz, jambalaya, and Zydeco music. It has given us Louis Armstrong and the Vampire Lestat. But most of all, New Orleans will always be the Mardi Gras—New Year's Eve, Halloween, and an R-rated Thanksgiving Parade all rolled into one. Mardi Gras is the big blowout that ends the season of Carnival, which dates back to the Latin words for "lifting up flesh" (or meat), and the revelry peaks on the day before Ash Wednesday, which opens the solemn Christian season of Lent. In New Orleans, Mardi Gras is a big, bawdy party that traces its roots to medieval France. It may go even farther back to an ancient Roman springtime festival called the Lupercalia. What else do you know about the Big Party and the Big Easy? Take this quick quiz.

1. What does "Mardi Gras" mean?

2. When did New Orleans join the United States?

3. In which war was the Battle of New Orleans fought?

4. Who are the "Cajuns"?

5. Who are the "Creoles"?

ANSWERS

1. It is from the French for "Fat Tuesday," and dates from a medieval French tradition of slaughtering a fatted bull or ox (*beouf gras*) on the day before Lent, a period in which eating meat was once forbidden.

2. The United States acquired New Orleans as part of the Louisiana Purchase in 1803. The state of Louisiana, and New Orleans, officially joined the Union in 1812.

3. In January 1815, General Andrew Jackson defeated the British there in the last battle of the War of 1812, even though that war had officially ended before the battle was fought. The victory propelled Jackson to national hero.

4. Derived from Acadia in Nova Scotia, Canada, "Cajun" is used to describe the descendants of thousands of French settlers who were forcibly transported to Louisiana by the British in 1755. That event was commemorated in Longfellow's famous poem, *Evangeline* (This is the forest primeval . . .).

5. Creoles are Louisiana descendants of early French and Spanish settlers.

DON'T KNOW MUCH ABOUT
Memorials

SINCE THE TERRIBLE EVENTS of 9/11—which some have taken to represent in Roman numerals as IX XI—there have been many notable memorials and remembrances, some organized, others highly impromptu. The plan for a permanent memorial at the World Trade Center site in New York, called "Reflecting Absence," was the result of a competition. The final plans for the memorial and development of the site are still being discussed and debated. In July 2006, The Port Authority of New York and New Jersey agreed to take over construction of the World Trade Center Museum and Memorial, which is scheduled to open on September 11, 2009, and is expected to cost at least $740 million. Of course, the idea of a memorial is an ancient one, and doesn't always honor the dead. Many American memorials are among the best known and most beloved sites in America. What do you know about some of the twenty-nine memorials currently operated by the National Park System?

1. Whose home is a memorial within Arlington Cemetery?

2. Completed in 1897, which memorial includes the tomb of a president and his wife?

3. Which site marks a tragedy that was one of the first successful Red Cross disaster relief efforts?

4. Opened in 1997, which memorial includes a collection of 168 symbolic empty chairs?

5. Which floating memorial spans a sunken ship?

Answers

1. Arlington House, home of Robert E. Lee, the Confederate general.

2. The General Grant National Memorial—referred to as Grant's Tomb—in New York City. When the memorial was opened, nearly 1 million people attended the dedication parade.

3. Pennsylvania's Johnstown Flood National Memorial, which commemorates the 2,209 people who died in the Flood of 1889.

4. The Oklahoma City national memorial, which commemorates the victims, survivors, and rescuers of the Oklahoma City bombing of April 19, 1995.

5. The U.S.S. *Arizona* memorial in Pearl Harbor, which attracts more than 1.5 million visitors each year.

HISTORIC
HAPPENINGS

DON'T KNOW MUCH ABOUT
the Emancipation Proclamation

ON JANUARY 1, 1863, Abraham Lincoln issued the famous document that gave him his nickname, "The Great Emancipator." As commander in chief during the Civil War, Lincoln called it "a fit and necessary war measure" because slave labor was helping the Confederacy continue its fight. What do you know about this essential document that eventually led to the end of America's "peculiar institution"?

TRUE OR FALSE?

1. The Emancipation Proclamation didn't really free any slaves.

2. Lincoln had tried to abolish slavery as soon as he was inaugurated in 1861.

3. Officially issued in January 1863, the Emancipation Proclamation was actually announced earlier.

4. Slavery did not end until the 15th Amendment was ratified.

5. The Proclamation also officially opened the Union military ranks to blacks.

ANSWERS

1. True. The Proclamation did not actually free a single slave, because it affected only areas under Confederate control and excluded slaves in the border states—Delaware, Kentucky, Maryland, and Missouri—and in other areas under Union control in Tennessee, Louisiana, and Virginia.

2. False. While Lincoln agreed that slavery was immoral, he believed it was still legal. Early in the war, he also felt that if he freed the slaves, he would divide the Union because the four slave-owning border states would secede.

3. True. On Sept. 22, 1862, five days after Union forces won the Battle of Antietam, Lincoln issued a preliminary proclamation stating that if the rebelling states did not return to the Union by January 1, 1863, he would declare their slaves to be "forever free."

4. False. The 13th Amendment, which became law on December 18, 1865, ended slavery in all parts of the United States.

5. True. The proclamation provided for the use of blacks in the Union Army and Navy, giving a crucial lift to the Union war effort. More than two hundred thousand black soldiers and sailors, including many former slaves, served in the armed forces at a time when Union enlistments were seriously declining.

DON'T KNOW MUCH ABOUT
the Lewis and Clark Expedition

ONE OF THE MOST extraordinary adventures in American history began more than two hundred years ago when the "Corps of Discovery" set off from St. Louis on May 14, 1804. Led by army veterans Meriwether Lewis and William Clark, the thirty-three permanent members of the expedition traveled over river and land through the unmapped stretches of wilderness to the Pacific coast. Returning with maps, descriptions of plants and animals, and information about the natives encountered, this remarkable quest opened up the vast riches of the northwest wilderness. What do you know about this incredible journey? Explore this quick quiz.

1. What were they looking for?

2. How far did they travel? How long did it take?

3. Who was Sacagawea?

4. Who was York?

5. How many of the men died?

6. What happened to Lewis?

ANSWERS

1. When Thomas Jefferson became president in 1801, he planned an expedition to discover a land and water route to the Pacific coast. With the purchase of the Louisiana Territory in 1803, the expedition grew to include mapping the new lands and laying claim to the Oregon region.

2. Their route took them up the Missouri River, across the Rocky Mountains, and along the Columbia and other rivers to the Pacific coast, and then back. The group traveled a total of about eight thousand miles, and returned to St. Louis in September 1806, nearly two and a half years after they left.

3. When the expedition reached the Mandan and Hidatsa Indian villages in what is now North Dakota, they established a winter camp at Fort Mandan and met a French-Canadian trader named Charbonneau who joined the expedition as guide. One of his wives, Sacagawea, a teenaged Shoshone Indian, also joined the expedition as an interpreter, carrying her newborn son. After her death, the boy, nicknamed Pomp, was raised by Clark.

4. William Clark's slave, he became the first black man to cross the continent north of Mexico. His features and strength were a source of awe to the Indians they met who had never seen white or black men.

5. Only one, apparently of a burst appendix.

6. The onetime private secretary to Thomas Jefferson, Lewis took his own life in 1809. Suffering from depression and alcoholism, he was not murdered as was rumored.

DON'T KNOW MUCH ABOUT
Watergate

IT WAS THE MOST FAMOUS "third rate burglary" in American history. On June 17, 1972, five men were arrested during a botched break-in at the Washington offices of the Democratic National Committee (DNC). For a time, it was a minor story as the nation reelected Richard Nixon, the Vietnam war dragged on, and Arab terrorists killed Israeli athletes at the Munich Olympics. But within a few months, the break-in was connected to the Nixon White House. Watergate, the "mother of all Gates," consumed the nation. What do you know about the criminal conspiracy that led to the indictments of more than forty government officials and brought down a powerful president? Try this quick quiz.

1. What were the "burglars" doing in Watergate?

2. Who were the "Watergate Seven"?

3. What was CREEP?

4. Who was "Deep Throat"?

5. How did Nixon leave office?

ANSWERS

1. The Watergate building housed an apartment complex and offices, including the office of the DNC. The "burglars" were trying to bug these offices.

2. The "Seven" included the five men arrested—four anti-Castro Cubans with ties to the CIA, and former FBI agent James W. McCord, along with two Republican political operatives, E. Howard Hunt, a former CIA agent involved in the Bay of Pigs operation, and G. Gordon Liddy, a former FBI agent. Tried and convicted, McCord disclosed that there was a much larger conspiracy and cover-up reaching into the White House.

3. The Committee to Re-Elect the President (Nixon). It was led by former attorney John Mitchell, who had ordered the break-in as part of a broader operation of buggings, break-ins, and political "dirty tricks" orchestrated by the White House and CREEP.

4. "Deep Throat" was the code name of the secret source of *Washington Post* reporters Bob Woodward and Carl Bernstein, whose investigative reporting kept the story on the front pages. Named for a notorious X-rated movie of that time, the identity of "Deep Throat" was revealed to be W. Mark Felt, former assistant director of the FBI during the Nixon administration. The long-held secret was finally disclosed by Felt in May 2005.

5. Named an unindicted "coconspirator" in the burglary, Nixon faced three articles of impeachment. After admitting his involvement in the cover-up and interfering with an FBI investigation, Nixon announced his resignation on August 8, 1974, before he could be impeached. On September 8, he was granted an unconditional pardon by his successor, President Gerald Ford.

DON'T KNOW MUCH ABOUT

the Nuremberg Trials

ON NOVEMBER 20, 1945, the first trials of German war criminals began at 10 A.M. in Nuremberg, Germany, once the site of massive Nazi rallies. Following Germany's defeat in World War II, this marked the first time in modern history that victors in a war conducted war crimes trials. In thirteen trials that lasted until 1949, many Nazi officials and high-ranking German military leaders were prosecuted for their part in the extermination of 6 million Jews and more than 5 million other Europeans. What do you know about this momentous event?

TRUE OR FALSE?

1. The opening trial was conducted by the United States.

2. Among the charges against the defendants were war crimes and crimes against humanity.

3. The Nuremberg trials marked the first time many people saw evidence of the Holocaust.

4. Of the first twenty-two defendants, all were convicted and executed.

5. The defense of "just following orders" was permitted.

ANSWERS

1. False. The International Military Tribunal conducted the first trial. It consisted of eight judges, two each from the principal allied countries—the United States, the United Kingdom, France, and the Soviet Union.

2. True. War crimes covered such acts as the murder of prisoners of war and the excessive destruction of land and cities. Crimes against humanity referred to deporting civilians and using them for slave labor; conducting inhumane medical experiments; and persecuting and murdering people because of political beliefs, race, or religion.

3. True. Film shot by allied liberators and eyewitness accounts of the camps provided grisly evidence never before shown publicly.

4. False. Of the initial twenty-two tried, nineteen were convicted and twelve received death sentences, including Hermann Goering, General Alfred Jodl, and Martin Bormann, who was convicted in absentia (while absent). His remains, found in Berlin, were identified in 1972. Goering committed suicide before his scheduled execution and ten men were hanged on October 16, 1946. Twelve later trials were conducted by U.S. judges. They involved 185 defendants, including Nazi doctors. More than half of these defendants were convicted.

5. False. Although the "superior orders" defense was considered as a mitigating factor, the trials rejected the argument that orders from superiors relieve people from responsibility for war crimes. The trials confirmed the idea that soldiers have a moral duty to disobey inhumane orders.

DON'T KNOW MUCH ABOUT
the St. Louis World's Fair

"MEET ME IN ST. LOUIS" were the famous words it inspired. It was called the Louisiana Purchase Exposition in celebration of the centenary of the Louisiana Purchase in 1803. But the St. Louis World's Fair, as it is known, missed by a year and opened on April 30, 1904. Covering twelve hundred acres, it was the biggest and grandest World's Fair ever, and has acquired legendary status. From the foods it introduced, to the international exhibits it included and its monumental scale, the St. Louis World's Fair was a wonder. What do you know about this landmark in American popular history? Take this quick quiz.

TRUE OR FALSE?

1. Among the notable firsts at St. Louis: the hot dog, iced tea, the ice-cream cone, air conditioning, automatic player pianos, and T-shirts.

2. Many of the fifteen hundred buildings from the fair are still in use.

3. The St. Louis Fair introduced the ferris wheel.

4. The first Olympic games in America were conducted during the fair.

5. Among the events was "Negro Day" in August.

ANSWERS

1. Mostly true, although there are some disputes over the ice-cream cone, which was patented in 1903.

2. False. Of the public buildings, only the Palace of Fine Arts was built to last. Today it is the St. Louis Art Museum. The fair's administration building is now part of Washington University.

3. False. Known as the Observation Wheel, it had debuted at Chicago's 1893 Columbian Exposition and was named for George Ferris, its designer.

4. True. Due to war between Japan and Russia, only twelve nations participated at the St. Louis Games. The first concrete stadium in the world was built for the games. One notable downside: sewage flowed into an artificial lake used for water events and three water polo players reportedly later died of typhoid.

5. True. Fair organizers had to respond to complaints that blacks were not hired during construction and were not welcome at its restaurants.

DON'T KNOW MUCH ABOUT
LBJ's "War on Poverty"

DURING HIS TIME in office in the 1960s, President Lyndon B. Johnson changed America with a round of social programs akin to the Depression Era's "New Deal." One of these was announced on January 8, 1964, in the State of the Union message when LBJ ambitiously promised, "This administration today, here and now, declares unconditional war on poverty in America," which then stood at 22 percent. Economists and historians are divided on the outcome of that war, but LBJ and his programs transformed the country. What do you know about this controversial president whose legacy is marked by two wars—one on poverty and one in Vietnam?

1. Why did JFK choose LBJ as a running mate?

2. What was the "Great Society"?

3. Besides the War on Poverty, what other legislation was part of the Great Society?

4. What was LBJ's big surprise in 1968?

5. What was the U.S. poverty rate in 2004?

ANSWERS

1. The most powerful man in the Senate, Texas Senator Johnson was chosen to woo southern voters by Kennedy, a Bostonian and a Roman Catholic. Johnson was more conservative and a southerner. In the November election, Kennedy and Johnson narrowly defeated the Republican team of Vice President Richard M. Nixon and Henry Cabot Lodge, Jr.

2. A catchall term for the many social programs LBJ created, the Great Society was first used by LBJ in a May 1964 speech at the University of Michigan. The Great Society's impact has been enormous and controversial. Defenders say it improved the lives of millions, while critics contend it created a large, wasteful government with too much control.

3. During 1964 and 1965, Great Society legislation created, among others, Medicare and Medicaid; federal aid to schools; created Head Start, a federal preschool program, which began in 1965 as a summer project. Great Society legislation also dealt with civil rights, and the Civil Rights Act of 1964 and Voting Rights Act of 1965 are considered two of the most powerful anti-discrimination laws ever passed.

4. With antiwar sentiment growing, Johnson shocked the nation by announcing in March that he would not run for reelection. At the same time, he announced a reduction in bombing of North Vietnam.

5. According to Census Bureau figures in 2005, there were 37 million people in poverty (12.7 percent) in 2004, up from 35.9 million (12.5 percent) in 2003. The poverty threshold for a family of four in 2004 was an income of $19,307; for a family of three, $15,067; for two, $12,334; and for unrelated individuals, $9,645.

DON'T KNOW MUCH ABOUT
the Burr-Hamilton Duel

ON JULY 11, 1804, two of America's most extraordinary characters squared off at ten paces in an "affair of honor." The pistol duel in Weehawken, New Jersey, that killed Alexander Hamilton and left Aaron Burr accused of murder was the final act of a bitter rivalry that changed history. Who were these men and why did they fight?

1. Who was Alexander Hamilton?

2. Who was Burr?

3. Why did they fight a duel?

4. Who shot first?

5. What became of Burr?

ANSWERS

1. An illegitimate child from the West Indies, Hamilton became George Washington's aide during the Revolution, a chief architect of the U.S. Constitution, and the nation's first Treasury Secretary. He had resigned that post over an affair with a married woman whose husband tried to blackmail Hamilton.

2. Heir of a prominent family, Burr was a Revolutionary War veteran turned New York politician. In 1800, Burr landed in a tie for the presidency and Hamilton helped engineer Jefferson's election in the House of Representatives. Burr became vice president and was still in office when the duel took place.

3. Hamilton and Burr were political rivals and disliked each other intensely. Hamilton had worked against Burr in the presidential election and then again in 1804 when Burr ran for governor of New York. In that campaign, public rumors about Burr's romantic affairs were attributed to Hamilton, which brought on Burr's challenge.

4. History is unclear. Hamilton's shot was way off and some say he deliberately missed; others believe that Burr fired first and hit Hamilton, whose shot went wild.

5. Indicted for murder, Burr was never tried and presided over the Senate until his term ended. After his vice presidency, Burr was tried for treason in 1807 over a scheme to form an empire in the southwest. He was acquitted. He later prospered as a New York attorney until his death in 1836.

DON'T KNOW MUCH ABOUT

D-Day

Steven Spielberg's World War II epic, *Saving Private Ryan*, brought the brutal reality of combat home to millions, but many moviegoers did not know which battle the film depicted, or when and why it happened. The assault, code-named Operation Overlord, occurred June 6, 1944, against Hitler's Germany. In the largest amphibian assault in history, allied armies crossed the English Channel to land on five beaches in Normandy in northern France. The invasion force involved 700 ships, 4,000 landing craft, 10,000 planes, and some 176,000 allied troops. How much do you know about D-Day?

True or False?

1. The allied invasion force included troops from all NATO members.

2. The D-Day invasion marked the first allied assault on the European mainland.

3. The allied forces were commanded by General Dwight D. Eisenhower.

4. Following D-Day, the war against Germany continued for almost a year.

ANSWERS

1. False. The North Atlantic Treaty Organization (NATO) was not formed until 1949. Most of the troops participating were American, British, and Canadian.

2. False. The allies began to retake Europe by invading Italy in 1943.

3. True. As Supreme Commander of the allied forces, Eisenhower had ultimate responsibility for the invasion.

4. True. The German army did not formally surrender until May 7, 1945. May 8, 1945 was declared V.E. (Victory in Europe) Day.

WHAT ELSE DO YOU KNOW about the day later known as "The Longest Day"? Here's a bonus quiz.

1. What does the "D" in D-Day mean?

2. Among the U.S. commanders in the landing was the relative of two American presidents. Who was he?

3. What were the code names of the five beaches assaulted that day?

4. How much longer did the war in Europe last after D-Day?

5. Where is America's D-Day Museum?

ANSWERS

1. According to U.S. Army manuals, D-day is the term for a secret or undetermined date when a military operation is to begin. The codes "H-hour" and "D-day" were first used during World War I.

2. General Theodore Roosevelt, Jr., Teddy's son and FDR's cousin. A brigadier general, he won the Medal of Honor for his actions in leading troops ashore. He died of a heart attack in France on July 12.

3. American forces landed at "Utah" and "Omaha," Canadians at "Juno," and British forces landed at "Sword" and "Gold."

4. Nearly a year. An official surrender was signed on May 7, 1945, and May 8, 1945 was declared V-E (Victory in Europe) Day.

5. The D-Day Museum, America's national WW II museum, opened on June 6, 2000 in New Orleans.

World War I

IT WAS ON "the eleventh hour of the eleventh day of the eleventh month" in 1918 that the bloody fighting of World War I finally came to an end. Once called the Great War, World War I involved more countries and caused greater destruction than any war until World War II. Set off by an obscure assassin and a web of military alliances, the war involved all of Europe's main powers. Trench warfare and gas attacks, air combat, modern battleships and submarines all contributed to a war that lasted four years and took the lives of nearly 10 million troops and possibly an equal number of civilians. Its aftermath transformed the modern world. What do you know about the "war to end all wars"?

1. Who was Archduke Francis Ferdinand?

2. Which were the chief nations involved on each side?

3. What other great event took place in Russia during the war?

4. What was the *Lusitania*?

5. When did the United States join the war?

Answers

1. A member of Austria-Hungary's royal family, he was gunned down on June 28, 1914, in Sarajevo by an assassin with ties to a terrorist organization in Serbia. The government of Austria-Hungary believed that Serbia was behind the assassination and used the incident to declare war on Serbia. His murder sparked the outbreak of World War I.

2. When the fighting began, France, Britain, and Czarist Russia, who were known as the Allies, backed Serbia. More than twenty countries, including Italy, the United States, and Japan, eventually joined the Allies. They were opposed by the Central Powers, made up of Austria-Hungary, Germany, and the Ottoman Empire, based in what is now Turkey.

3. The Russian Revolution. Germany helped the Russian revolutionary Lenin, then living in Switzerland, to return home in April 1917. Seven months later, Lenin led an uprising that gained control of Russia's government and he immediately called for peace talks with Germany, taking Russia out of the war.

4. A British passenger liner secretly carrying arms, the *Lusitania* was torpedoed on May 7, 1915 by a German submarine, or U-boat. There were 128 Americans among the 1,198 passengers who died. The sinking of the *Lusitania* led President Woodrow Wilson to urge Germany to give up unrestricted submarine warfare, but it was not the reason America joined the fighting.

5. April 6, 1917. President Wilson had tried to keep America neutral and even campaigned with the slogan "He Kept Us Out of War" in the 1916 election. Tension between America and Germany increased after the British passed on a secret message that revealed a German plot to persuade Mexico to go to war against the United States. After U-boats sank several U.S. cargo ships, the war fever grew.

DON'T KNOW MUCH ABOUT

the Arab Oil Embargo

ON OCTOBER 17, 1973, America woke up to a new reality. During an Arab-Israeli war, the major oil-producing Arab states turned off the taps and $3-a-barrel oil was history. For the first time since the establishment of OPEC in 1960, these small countries flexed their collective muscles and Americans knew how much they relied on foreign oil. This oil embargo and subsequent runup in fuel prices had America waiting in long lines at the pumps and bundling up in sweaters, altered domestic politics, and damaged the American economy for years to come. What do you know about this crucial modern event?

1. What was the "Yom Kippur War"?

2. How did American filling stations ration gasoline?

3. What major change in federal law was passed as a result of the boycott?

4. How long did the embargo last?

5. What does OPEC mean?

6. How many current OPEC members are not Arab states?

ANSWERS

1. Also called the October War and the Ramadan War, it was launched by Egypt and Syria against Israel on October 6, 1973 and lasted until a U.N. ceasefire was signed on October 22. Both sides suffered heavy casualties with no clear-cut victor. American aid to Israel led to the oil embargo.

2. With supplies short, stations filled tanks based on license plate numbers; licenses ending in even or odd numbers could get gasoline on alternating days. Drivers could only get gasoline if they had less than half a tankful.

3. Congress imposed the 55 mph speed limit in 1974 as a measure to improve the fuel efficiency of cars during the Arab oil embargo. It has since been raised.

4. It lasted until March 18, 1974. By the end of 1973, oil prices had risen from $3.00 to $11.65 per barrel, peaking in the mid–1970s at around $40 per barrel.

5. It stands for the Organization of Petroleum Exporting Countries, an eleven-member group that attempts to set world oil prices by controlling production. OPEC's members now collectively supply about 40 percent of the world's oil output, and possess more than three-quarters of the world's total proven crude oil reserves. Although not all members of OPEC are Arab states, the Arab countries held much of OPEC's power in 1973.

6. The current non-Arab OPEC members are Indonesia, Nigeria, Venezuela, and Iran. Algeria, Iraq, Kuwait, Libya, Qatar, Saudi Arabia, and the United Arab Emirates are the other members.

DON'T KNOW MUCH ABOUT
the Korean War

IT STARTED IN 1950 and, as the headlines continued to prove, it never really ended. But fighting in the Korean War, one of the bloodiest wars in history, ceased on July 27, 1953, when the U.N. and North Korea signed an armistice. A permanent peace treaty between South Korea and North Korea has never been signed. What else do you know about this Cold War conflict that had the world on the brink of World War III?

1. When did the Korean War begin?

2. Who commanded the U.N. troops in Korea?

3. What nation entered the war on North Korea's side?

4. What two aviation "firsts" occurred during the war?

5. Why did President Truman fire General MacArthur?

6. What were American losses in the Korean War?

ANSWERS

1. June 25, 1950, when troops from Communist-ruled North Korea invaded South Korea. The U.N. called the invasion a violation of international peace and demanded that the Communists withdraw. Sixteen U.N. countries sent troops to help the South Koreans, and forty-one countries sent military equipment and other supplies. The United States provided about 90 percent of the troops, military equipment, and supplies.

2. On July 8, with the approval of the U.N. Security Council, President Truman named Douglas MacArthur commander in chief of the United Nations Command.

3. More than three hundred thousand Chinese troops crossed into North Korea in October 1950 and U.S. and Chinese troops first clashed on October 25. They fought until November 6, when the Chinese suddenly withdrew.

4. The Korean War marked the first battles between jet aircraft and, for the first time, helicopters carried troops into combat.

5. One of the controversies of the war occurred in April 1951, when President Truman removed General MacArthur from command, the result of a continuing dispute between MacArthur and defense leaders in Washington. MacArthur wanted to bomb bases in a part of China, and use other "all-out measures." Truman feared such actions might lead to a third world war.

6. The Department of Defense reports that 54,246 American service men and women lost their lives during the Korean War. This includes all losses worldwide during that period. As there has been no peace treaty, those Americans who lost their lives in the Demilitarized Zone of Korea since the Armistice are also included.

DON'T KNOW MUCH ABOUT
the Civil Rights Movement

FULL OF TRAGEDY and triumph, 1963 was one of the most eventful years in American history, a year of turning points in the nation's quest for racial equality. It was the year that President Kennedy declared that the struggle for civil rights was a "moral issue." It was the year that saw the murder of civil rights activist Medgar Evers. What else do you know about this landmark year in the nation's struggle for civil rights?

1. At his inauguration on January 14, 1963, who said, "Segregation now, segregation tomorrow, and segregation forever."

2. In what city did a peaceful march turn violent?

3. What historic event took place on August 28, 1963?

4. What event in Birmingham, Alabama, shocked the nation on September 15, 1963?

5. Which black actor achieved a historic first in 1963?

6. What first did New York Yankees baseball great Elston Howard achieve in 1963?

7. Which service academy graduated its first blacks in 1963?

ANSWERS

1. Alabama Governor George Wallace who would later block the entrance to the University of Alabama to the first black students admitted there.

2. A march in Birmingham, Alabama, led by Martin Luther King and Ralph Abernathy, turned ugly when police used attack dogs and fire hoses to disrupt the protest.

3. The famed March on Washington drew more than 250,000 Americans to the steps of the Lincoln Memorial where they heard Martin Luther King's immortal "I Have a Dream" speech.

4. A bomb exploded in a Baptist church killing four young black girls attending Sunday school.

5. Sidney Poitier became the first black to win the Academy Award for Best Actor for *Lilies of the Field*.

6. Catcher Howard, the first black Yankee, became the first black to win the American League's Most Valuable Player award.

7. Three blacks who had entered the Air Force Academy in 1959 graduated in 1963.

DON'T KNOW MUCH ABOUT

the Salem Witch Trials

THE MOVIE SENSATION, *The Blair Witch Project*, served up a chilling reminder of how much we love to hate witches. Setting aside the image of Samantha from "Bewitched," the notion of a warts-and-broomstick hag with high-peaked hat has been around for a long time, usually inspiring more dread than Halloween fun. But once upon a time, witchcraft in America was no laughing matter. In the Massachusetts Bay Colony in 1692, a panic over witchcraft threw the village of Salem into turmoil, with tragic results. What do you know about the witches of Salem?

TRUE OR FALSE?

1. The Salem "witch hunt" began when a Carib Indian slave named Tituba accused townspeople of being witches.

2. Once condemned, a "witch" could avoid execution by confessing.

3. Although dozens of people were condemned as witches, none were executed.

4. The famous courtroom drama *Inherit the Wind* was based on the Salem Witch Trials.

Answers

1. False. The incident started when two young girls began acting strangely. They accused Tituba, the slave of the town's minister, of teaching them witchcraft. She in turn testified that other townspeople were part of a witchcraft conspiracy.

2. True. Only those who continued to proclaim their innocence were executed.

3. False. Nineteen people were hung as witches and another man was pressed to death under stones for refusing to be tried.

4. False. Arthur Miller's play *The Crucible* is based on the Salem trials. *Inherit the Wind* is based on the Scopes Monkey Trial.

DON'T KNOW MUCH ABOUT
the Great Crash

T.S. ELIOT ONCE SAID, "April is the cruelest month." But to many Wall Street investors, the real cruelest month is October—the month that conjures up visions of Wall Street's darkest days. In 1929, the losses on October 28 ("Black Monday") and October 29 ("Black Tuesday") mark these as dates that live in infamy for stock market historians and investors with long memories. After years of wild speculation during the "Roaring 20s," the Great Crash of 1929 wiped out billions of dollars and helped usher in the Great Depression, an era of widespread bankruptcy, severe unemployment, and tremendous hardship. It was a different Wall Street back then. Many fewer companies were traded and there was little regulation of the investment business. What else do you know about the days that still make an investor's blood run cold?

1. How big was the fall on October 29, 1929?

2. How many shares were traded that day?

3. How long did it take for the stock market to regain its previous high of 381?

4. Which companies in the Dow Jones Industrial Average on October 29, 1929 are still in the Dow 30?

5. What government agency was created to monitor the stock market? Who was the first man in charge of it?

6. Was Black Tuesday the largest one-day loss in percentage in Dow Jones history?

ANSWERS

1. The Dow Jones Industrial Average lost 30.57 points that day, closing at 230, a percentage loss of 11.73 percent. The preceding day had actually been worse and in the two days combined, the stock market had fallen nearly 40 percent from its peak. With the Industrial Average above 13,000 in Spring 2007, a similar decline would take the DJIA back to 7,800!

2. A record 16.4 million shares were traded on Black Tuesday. Today, a billion shares commonly change hands and the New York Stock Exchange can accommodate trading volume of more than 2 billion shares.

3. Twenty-five years. The high of 381 was not reached again until November 24, 1954.

4. Only two. General Electric, one of the original Dow 12, and General Motors.

5. The Securities and Exchange Commission (SEC) was created by Congress in 1934 to regulate commerce in stocks and bonds. Its first commissioner was Joseph P. Kennedy, father of President John F. Kennedy.

6. No. The largest one-day percentage loss came on October 19, 1987. That day's loss of 508 points was a 22.61 percent decline.

DON'T KNOW MUCH ABOUT
the Civil War

THE CONTROVERSY OVER the Confederate flag flying in several southern states is the latest reminder that the wounds of the Civil War have not fully healed more than 140 years after the end of the war that killed an estimated six hundred thousand Americans. Between 1861 and 1865, many of the war's most significant events occurred during April. What do you know about April's prominence in the Civil War?

1. On April 12, 1861, what federal fort was attacked by Confederate forces?

2. On April 19, 1861, civilians and Union soldiers became the war's first casualties during a riot in which Union city?

3. On April 7, 1862, one of the bloodiest battles of the war was fought at Pittsburg Landing, Tennessee, a battle best known by what name?

4. On April 16, 1862, Confederate President Jefferson Davis signed a law creating an American first. What was it?

5. On April 25, 1862, what major Confederate city surrendered to Union Admiral Farragut?

6. What atrocity occurred on April 12, 1864?

7. Where did Robert E. Lee surrender to Ulysses S. Grant on April 9, 1865?

8. What event shook the nation on Good Friday, April 14, 1865?

ANSWERS

1. The assault on Fort Sumter in the harbor of Charleston, South Carolina, is considered the beginning of the Civil War.

2. Baltimore, Maryland, where Union soldiers marching to Washington were attacked by a secessionist mob.

3. Shiloh.

4. The Conscription Act, the first American military draft.

5. New Orleans.

6. The massacre of surrendering black soldiers at Fort Pillow in Tennessee.

7. Appomattox Courthouse, Virginia.

8. Lincoln was shot. He died the next day.

HOLIDAYS
AND
TRADITIONS

DON'T KNOW MUCH ABOUT
New Year's Eve

CELEBRATING A NEW YEAR is an old idea. Just about every ancient civilization did it, including the Babylonians. One night wasn't good enough for them; theirs lasted eleven days. Many other cultures celebrated a New Year rite, but they often came in the springtime when a new planting season began. The January date started in ancient Rome, but the New Year kept moving until the Gregorian calendar was set by the Pope in 1582. And yes, Dick Clark was there. Remember an "auld acquaintance" and resolve to try this quick quiz.

1. What is "Auld Lang Syne" anyway?

2. Which famous building celebrated its opening on New Year's Eve in 1904?

3. When did the Rose Parade begin?

4. What Dutch treat is popular for the New Year?

ANSWERS

1. An old Scottish tune, it means "old long ago," or "the good old days." Sung in most English-speaking countries at midnight, the traditional song was set down in its familiar form by Robert Burns and first published in 1796.

2. The *New York Times* building celebrated its opening with fireworks on New Year's Eve,1904, and Times Square has been the place to be ever since. The lowering of the ball tradition came a few years later.

3. In 1886, to celebrate the orange crop. Football was played in 1902 but was replaced the next year by chariot races! The Rose Bowl game became a permanent fixture in 1916.

4. Doughnuts. Many cultures use ring-shaped food to symbolize the year coming "full circle."

DON'T KNOW MUCH ABOUT
Easter Customs

JELLY BEANS, marshmallow bunnies, and chocolate eggs. Easter Sunday, like Christmas, is an important religious holiday that has become increasingly secular and commercial. But around the world, Christians celebrate the resurrection of Jesus in very similar ways with painted eggs and baskets. What do you know about the most significant day on the Christian calendar?

TRUE OR FALSE?

1. The date of Easter moves with the Jewish holiday of Passover.

2. The word "Easter" comes from pagan mythology.

3. In Greece, Easter eggs are all red.

4. Teddy Roosevelt began the White House Easter egg roll tradition with his children.

5. The author of the popular song "Easter Parade" was Jewish.

6. Hot cross buns on Good Friday are a British invention.

ANSWERS

1. False. Early Christians tied the celebration of Easter to Passover, but Easter's date for western Christians is now tied to the first Sunday after the first full moon that occurs on or after the vernal equinox. This means Easter can fall between March 22 and April 25. Eastern Orthodox Christians use a different calculation based on the Julian calendar.

2. True. Eostre was the Anglo-Saxon fertility goddess. Many of the symbols of Easter, like rabbits, are ancient pagan fertility symbols. In most other European countries, the word for Easter is closely related to the Jewish word *Peach* for Passover.

3. True. Greek Easter eggs are painted red for the blood of Jesus Christ. The tradition of painted eggs is another ancient fertility symbol.

4. False. It was begun by Rutherford B. Hayes's First Lady Lucy Hayes in 1878. Prior to that, there had been egg rolling at the U.S. Capitol.

5. True. Russian-born Israel Balline became famous as Irving Berlin and wrote the song "Easter Parade" in 1933. It inspired the 1947 Fred Astaire-Judy Garland film of that name. New York's tradition of people promenading in their Easter finery began during the Civil War era.

6. False. While popular in England, the custom of hot cross buns, which have a cross symbolizing the crucifix, goes back to pre-Christian times when Zeus was offered a bun with a cross supposed to represent a bull's horns. Later Christians believed that hot cross buns had miraculous healing powers.

DON'T KNOW MUCH ABOUT
Passover

EVEN IF YOU'VE SEEN *The Ten Commandments* every year it's been shown, you may not connect it with Passover. This Jewish festival celebrates the flight of the Israelites from Egyptian slavery, traditionally believed to have happened in 1200s B.C. Like many Jewish holidays, Passover is tied to the lunar calendar and usually begins in March or April. What else do you know about the most widely observed Jewish holiday? Find out why "this night is different from all others" in this quick quiz.

TRUE OR FALSE?

1. The Passover story is told in the book of Genesis.

2. The word "Passover" comes from the biblical story of the plagues which God brought on Egypt.

3. Passover is celebrated with the ceremonial lighting of candles on eight nights.

4. The most important symbol of the Passover is unleavened bread called matzo.

5. The Last Supper of Jesus was a Passover seder.

ANSWERS

1. False. The story of Passover is told in Chapter 12 of the Book of Exodus.

2. True. As the tenth plague in the story, God killed the first-born child in every Egyptian home but would "pass over" (*pesach* in Hebrew) the homes of the Israelites. The word *Passover* also refers to the passing over of the Israelites from slavery to freedom.

3. False. The Passover is celebrated for seven days and centers on the Seder, a ceremonial meal when the story of the flight of the Israelites is read from a book called the *Haggadah*. During the ceremony, the youngest person asks, "Why is this night different from all others?" The celebration of lights is known as Hanukkah.

4. True. In the Exodus story, when the Israelites fled, they did not have time to let their bread rise so they made flat, unleavened bread instead. Therefore, Jews eat matzo (also spelled *matzoh*) instead of leavened bread during Passover.

5. True. Although not all scholars agree, three of the four gospels describe the Last Supper as a Passover meal. Christians in many European countries call Easter *Pascha*, which comes from the Hebrew word *pesach*.

DON'T KNOW MUCH ABOUT
Kwanzaa

GROWING IN POPULARITY and recognition each year, Kwanzaa is an African-American holiday that has come to be seen as a Christmastime celebration. But it is really more like Thanksgiving. Beginning on December 26 and lasting for seven days, Kwanzaa is drawn from traditional African harvest festivals. Developed nearly forty years ago in the U.S. by Maulana Karenga, a professor of Pan-African studies, it is a cultural rather than religious celebration that combines traditional African practices with African-American aspirations and ideals. What else do you know about this tribute to America's African past?

1. What does the word "Kwanzaa" mean?

2. What do the seven candles lit for each day of Kwanzaa mean?

3. The traditional colors of Kwanzaa are black, red, and green: What do they represent?

4. How does Kwanzaa end?

ANSWERS

1. The word *Kwanzaa* comes from a phrase that means "first fruits" in Swahili, the most widely spoken African language.

2. The holiday centers around the Nguzo Saba, seven principles of black culture that were developed by Professor Karenga. These principles are unity (Umoja), self-determination (Kuji-chagulia), collective work and responsibility (Ujima), coopera-tive economics (Ujamaa), purpose (Nia), creativity (Kuumba), and faith (Imani). Families light one of the seven candles in a *kinara* (candleholder) and discuss the principle for the day.

3. The colors of the Kwanzaa flag are black, red, and green; black for the people, red for their struggle, and green for the future and hope that comes from their struggle.

4. Near the end of the holiday, the community gathers for a feast called the Karamu, which is held on December 31. A typical karamu features traditional African music, dancing, and food, and also serves as a new year celebration.

DON'T KNOW MUCH ABOUT
American Christmas Customs

IF ALL YOU WANT for Christmas is a six-week holiday, move Down Under. Christmas is, of course, one of the world's most widely celebrated holidays, and those celebrations can be very different from the traditional American idea of Santa and snow. In fact, Christmas in Australia comes during the summer, so kids get a six-week "summer" vacation and an Aussie Christmas dinner might be served on the beach! What else do you know about Christmas customs around the globe?

1. What is Boxing Day?

2. Where do children put out their shoes instead of hanging stockings?

3. Where do bad children risk a visit from Father Spanker?

4. Where would you be served an eel on Christmas Eve?

5. Where does Christmas season begin on December 13?

ANSWERS

1. Celebrated on December 26 in England, Ireland, and Australia, the holiday goes back to the notion that noblemen "boxed up" gifts for their servants on this day. It is also called St. Stephen's Day.

2. During the evening of January 5, Spanish children put their shoes near a window. The next day is Epiphany, celebrating the visit of the Magi to the infant Jesus. According to legend, the Wise Men arrive during the night before Epiphany and fill the children's shoes with small gifts.

3. On Christmas Eve, French children put their shoes in front of the fireplace hoping that Father Christmas will come. But his partner, Father Spanker (Le Pere Fouettard) delivers a spanking to naughty children.

4. In Italy, that is a traditional Christmas Eve meal. Italian children receive gifts from La Befana, a kindly old witch, on the eve of Epiphany. According to legend, the Wise Men asked the kindly old witch to accompany them to see the infant Jesus.

5. In Sweden festivities begin on St. Lucia Day, December 13. In the morning of this day, the oldest daughter in the home dresses in white and wears a wreath with seven lighted candles on her head. She serves the other members of the family coffee and buns in bed.

DON'T KNOW MUCH ABOUT
Christmas Traditions

THERE IS NO BETTER example of America, the Melting Pot America, than our Christmas traditions. Almost every group has added a little something to the American idea of Christmas cheer, although no one wants to take credit for the fruitcake. Think you know the origins of some popular American Christmas traditions? Test your CQ (Christmas Quotient) in this quiz.

1. Where did the singing, candle-lighting, and gift-giving come from?

2. What is the source of "Xmas?"

3. Who brought the Christmas tree to America?

4. What country gave us the poinsettia?

5. Was Christmas ever outlawed in America?

6. Who thought up Christmas cards?

ANSWERS

1. The celebration of Jesus' birth date was influenced by pagan festivals in ancient Rome where year-end celebrations honored Saturn, their harvest god, and Mithras, the god of light. These celebrations gradually became part of Christmas custom for early Christians.

2. The word "Xmas," sometimes used instead of Christmas, also began in the early Christian church. In Greek, "X" is the first letter of Christ's name and was frequently used as a holy symbol.

3. The first Christmas trees in America were used in the early 1800s by German settlers in Pennsylvania. Evergreens were ancient pagan symbols of the renewal of life, but the Christmas tree probably developed in part from the "Paradise Tree," an evergreen decorated with apples used in a popular play about Adam and Eve held on December 24 in medieval Germany.

4. Mexico. Introduced to America by Dr. Joel Poinsett, the U.S. Ambassador to Mexico, the plant's red and green leaves and resemblance to a star made the poinsettia a popular Christmas decoration.

5. Yes, in Boston from 1659 to 1681. During the Protestant Reformation, many Puritans considered Christmas a pagan celebration because it included nonreligious customs. It was also thought to be too Catholic.

6. The first Christmas card was created in 1843 by English illustrator John Calcott Horsley. The first Christmas cards manufactured in the United States were made in 1875 by Louis Prang, a German-born Boston printer.

DON'T KNOW MUCH ABOUT
Presidents' Day

DID YOU KNOW George Washington had two birthdays? He was born on February 11, but the calendar was altered in 1752, so his "official" birthday became February 22. Abraham Lincoln was born on February 12. But then Congress decided to lump these two great American presidents together to make one Presidents' Day, now a movable feast that is observed on the third Monday in February. You may know who said "Fourscore and seven years ago" and "I cannot tell a lie." But can you tell George and Abraham apart by their words? Figure out who said it in this quick quiz.

MATCH THE QUOTE TO THE
CORRECT PRESIDENT.

1. "There is not a man living who wishes more sincerely than I do, to see a plan adopted for the abolition of [slavery]."

2. "If destruction be our lot we must ourselves be its author and finisher. As a nation of freemen we must live through all time, or die by suicide."

3. "No man is good enough to govern another man without that other's consent."

4. "The fate of unborn millions will now depend, under God, on the courage and conduct of this army."

5. "Let us have faith that right makes might, and in that faith let us to the end dare to do our duty as we understand it."

6. "It is well, I die hard, but I am not afraid to go."

ANSWERS

1. Washington to Robert Morris, 1786. Despite his misgivings, Washington kept slaves throughout his life.

2. Lincoln to Young Men's Lyceum, January 1838.

3. Lincoln, in an Illinois speech, October 1854.

4. Washington, to the Continental Army, August 1776.

5. Lincoln, address at Cooper Union, February 1860.

6. Washington's last words, December 14, 1799.

DON'T KNOW MUCH ABOUT
Irish-American History

MARCH 17 IS THE DAY that everyone good-naturedly claims some Irish blood. St. Patrick's Day honors a fifth-century Briton, once the slave of an Irish chieftain, who returned to Ireland as a missionary and converted the island to Christianity around A.D. 432. But in the midst of the whimsical references to four-leaf clovers and leprechauns, many people are unaware of a time in America when the sign "No Irish Need Apply" was a common sight. For many years, Irish Americans endured discrimination, derision, and prejudice. Is your Irish-American knowledge all "blarney"?

TRUE OR FALSE?

1. Beginning in 1845, some 2 million Irish came to America to escape religious persecution.

2. With its distinctive Irish harp flag and Irish language motto, *Riamh Nar Dhruid O Sapirn lann*, the mostly Irish 69th New York regiment won fame in the Civil War as the "Fighting 69th."

3. The "Molly Maguires" was a group of Irish nurses who served heroically during World War II.

4. The "Know Nothing" party was a nineteenth-century political party opposed to Irish immigration.

ANSWERS

1. False. The potato blight, a fungus which first struck Ireland in 1845, resulted in a major famine, sparking a wave of Irish immigration to America. During the famine, another million Irish people died of starvation, even as British landowners sent their Irish-grown food and livestock back to England for sale.

2. True. The Fighting 69th was involved in almost every major Civil War engagement and their nickname was given by opposing General Robert E. Lee. The Irish motto inscribed on the flag says, "Who never retreated from the clash of spears."

3. False. They were a group of mostly Irish coal miners in Pennsylvania in the late nineteenth century. They were accused of violence against mine owners, and, on dubious evidence, twenty-four men were convicted and ten of them were hanged.

4. True. This "Nativist," anti-immigrant, anti-Catholic political party formed in the 1840s, out of fear of the growing numbers of Irish Catholics who were swelling the ranks of the Democratic party. Some of its members were absorbed into the Republican party when it formed in the 1850s.

You may know that John F. Kennedy was an Irish-American president. But what else do you know about some great Irish-American achievers? There's no blarney in this bonus quiz.

1. What military hero, born to an immigrant farming family in the hills of Carolina, became the first American president of Irish descent?

2. What Irish American was the only Catholic to sign the Declaration of Independence?

3. Born to a pioneer family in Tennessee, who became one of the heroes of the Alamo?

4. Name two famous Irish-American senators with the same last name but very different political views.

5. Name America's first Nobel prizewinning playwright.

ANSWERS

1. Andrew Jackson, the seventh president. All of his predecessors were of British ancestry.

2. Charles Carroll III, born in Maryland. Despite strict anti-Catholic laws, Carroll served as a U.S. senator and was reputed to be America's richest man when he died in 1832.

3. Davy Crockett who spent some time in politics before moving to Texas where he died in 1836.

4. Wisconsin's Senator Joseph McCarthy, who started the anti-Communist purge of the 1950s, and Senator Eugene McCarthy of Minnesota who ran for president in 1968 as a peace activist.

5. Eugene O'Neill, son of an immigrant family, who wrote such masterpieces as *The Iceman Cometh, Long Day's Journey into Night,* and *A Moon for the Misbegotten.*

DON'T KNOW MUCH ABOUT
Independence Day

IT IS A MERE 1,337 words, most of them a laundry list of complaints against the British king. But more than 230 years after the Continental Congress adopted Thomas Jefferson's Declaration of Independence in Philadelphia on July 4, 1776, this handful of words remains one of history's most powerful statements. When John Hancock attached his famous signature to the declaration, he urged a unanimous vote: "There must be no pulling different ways. We must hang together." The seventy-year-old Ben Franklin quickly agreed, "We must indeed all hang together, or most assuredly we shall all hang separately." What do you know about the document that told the world that the United States of America was "free and Independent"?

1. When did the Continental Congress actually pass a resolution of Independence?

2. Who was on the committee assembled to draft the declaration?

3. Why did Adams tell Jefferson, the youngest man on the committee, to draft a declaration?

4. What did Congress cut out of Jefferson's draft of the declaration?

5. Which two key players in the history of Independence died on the fiftieth anniversary of the declaration?

ANSWERS

1. On July 2, the Lee-Adams resolution of independence was adopted and many thought that would be the date celebrated as America's birthday.

2. John Adams, Ben Franklin, Robert Livingston of New York, and Roger Sherman of Connecticut. Jefferson was selected because Virginia was so politically powerful.

3. First, the thirty-three-year-old Jefferson was a Virginian; second, Adams knew that he himself was unpopular and considered obnoxious by many delegates. Finally, Adams said Jefferson could "write ten times better than I can."

4. The Congress made 86 changes, which eliminated 480 of Jefferson's words. Most important was the removal of all references to slavery, "the execrable commerce," which Jefferson, a slaveholder himself, had blamed on King George.

5. Thomas Jefferson and John Adams both died within hours of each other on July 4, 1826.

DON'T KNOW MUCH ABOUT
the Chinese New Year

UNLIKE JANUARY 1 in the western calendar, Chinese New Year's Day is a movable feast that is fixed to the new moon of the first month in the Chinese lunar calendar. Each year the celebration of Chinese New Year involves fireworks, drums, and noisy gongs. That tradition comes from an ancient legend in which people used loud noises to drive an evil monster away from their village. What else do you know about this nearly five-thousand-year-old tradition? Take this quick quiz.

TRUE OR FALSE?

1. The Chinese new year greeting "Gung Hey Fat Choy" means "May you live one hundred years."

2. Each Chinese new year corresponds to one of the nine animals in the Chinese zodiac.

3. In the Chinese calendar, some years get an extra month—a leap month.

4. Gunpowder was developed by the Chinese for firecrackers about two thousand years ago.

ANSWERS

1. False. It means "Wishing You Prosperity and Wealth."

2. False. There are twelve animals in the Chinese zodiac: rat, ox, tiger, hare, dragon, snake, horse, sheep (or goat), monkey, rooster, dog, and pig (or boar).

3. True. Because the Chinese lunar year is shorter than 365 days, the calendar is adjusted by inserting an extra month every seven years in a nineteen-year cycle. In 2006 it was a Chinese Leap Year.

4. False. Gunpowder was developed in China in the ninth century.

DON'T KNOW MUCH ABOUT
Valentine's Day

LIKE MANY FAMILIAR holidays, Valentine's Day and its origins are shrouded in myth. Some trace its beginnings to an ancient Roman festival, the Lupercalis, which was celebrated on February 15. But recent research attributes the idea of Valentine's Day to the English poet Chaucer who popularized the holiday in a 1380 verse, *Parlement of Foules*. Despite the myths, one thing is certain about this day devoted to lovers: it sells a lot of cards, chocolates, flowers, and lingerie. And woe to those who forget. An arrow in the heart might be less lethal than a sweetheart scorned. Have the heart to try a Valentine's Day quiz?

1. Was there really a St. Valentine?

2. The Romans called him Cupid and he is the symbol of the day. What did the Greeks call the little god of love?

3. In 1828, Dutchman Conrad Van Houten changed romantic history when he created what Valentine staple?

4. Who is considered the most romantic movie idol of all time?

Answers

1. Yes. Actually there were two. A third-century Valentine was martyred during the persecution by Rome's Claudius II. Another martyr named Valentine was an early bishop and has also been suggested as the inspiration of the modern St. Valentine's Day. The feast of St. Valentine was marked on February 14.

2. Eros, which means "sexual love" in Greek, was the son of Aphrodite and was represented as a handsome youth. In Roman myth, Cupid was the son of Venus and was usually depicted as a chubby infant. In Latin, *cupido* means desire.

3. The chocolate bar. Cocoa, introduced to Europe in 1527 by Spanish conquistadors, was considered an aphrodisiac and was consumed in liquid form until Van Houten's marvelous discovery.

4. Rudolph Valentino, the silent movie actor whose death in 1926 brought one hundred thousand mourners to his funeral and left millions of women grieving.

DON'T KNOW MUCH ABOUT

Memorial Day

ALTHOUGH MANY OF US are primed to kick off the summer's first picnics and barbecues on the Memorial Day weekend, the holiday is one of the most solemn events on the American calendar. Designated to honor America's war dead, it recognizes the supreme sacrifice made by men and women throughout American history.

1. What was the original name for Memorial Day?

2. When was it first celebrated?

3. What was America's most deadly war?

ANSWERS

1. Memorial Day began as Decoration Day, a national day of remembrance on which the graves of dead soldiers were adorned with flowers.

2. The Decoration Day tradition began in 1868, following the Civil War. Originally celebrated on May 30, the date was chosen by General John Logan, commander of the Grand Army of the Republic. It was meant as a national day of remembrance to replace local decoration days. In 1971, Congress set Memorial Day on the last Monday in May. Several states still honor the original date. Six states also recognize separate Confederate memorial days.

3. The Civil War. An estimated 620,000 Americans died fighting in the Civil War, approximately 2 percent of the population at the time. By contrast, some 403,000 American servicemen and women died in World War II.

DON'T KNOW MUCH ABOUT
Santa Claus

IN SOME PLACES, he is Father Christmas, Kris Kringle, or Pere Noel. But here in America, the legendary old man who brings gifts to children at Christmas is Santa Claus. Long before Christianity was founded, the custom of giving gifts on a special day in winter was practiced. Once Christianity was established, the idea was gradually attached to Christmas in recognition of the gifts given by the Wise Men to Jesus. Saint Nicholas, whose feast day is December 6, gradually became a symbol of this custom among Christians. During the Reformation of the 1500s, Protestants began to substitute nonreligious characters for Saint Nicholas, who eventually evolved into the character now known as Santa Claus. Know your Santa legends? Naughty or nice, try this quick quiz.

TRUE OR FALSE?

1. Santa Claus is based on an actual person.

2. The concept of Santa Claus was brought to America by early British settlers.

3. Saint Nicholas was once depicted as tall and thin, wearing a bishop's robe, and riding a white horse.

4. In other countries, there are many other imaginary characters who bring gifts on a certain day of the year other than Christmas.

ANSWERS

1. Maybe. Saint Nicholas, a legendary Christian figure, from around A.D. 300, was believed born in what is now Turkey. A legend grew that he once aided a poor man whose daughters had no dowry. Nicholas threw bags of money through an open window of the man's house and the daughters were able to marry. The idea of Saint Nicholas as a man who brings gifts grew out of this tale.

2. False. The first Dutch settlers who came to America had a figure of Saint Nicholas on the front of their ship and the Dutch maintained the custom of celebrating the saint's feast day on December 6. Over time, English settlers in America adopted these Dutch legends and festivities associated with Saint Nicholas but his Dutch name, Sinterklaas, was soon turned into Santa Claus.

3. True. In 1809, American writer Washington Irving presented a new image of the saint. He described Saint Nicholas as a stout, jolly man who wore a broad-brimmed hat and smoked a long pipe. Irving's Saint Nicholas rode over treetops in a wagon filled with presents.

4. True. In the Netherlands and Belgium, Saint Nicholas visits homes on Saint Nicholas Eve, December 5, accompanied by a character named Black Pete, who carries a birch rod to whip naughty children. In southern Germany, people usually say the Christkind (Christ child) sends the gifts on Christmas Eve. From the name Christkind came the character Kris Kringle, who gradually became identified with Santa Claus. In France, Pere Noel leaves small presents in homes on Christmas Eve. In Sweden, Denmark, and Norway, an elflike character brings gifts on Christmas Eve.

EVERYDAY
OBJECTS
AND
REMARKABLE
INVENTIONS

DON'T KNOW MUCH ABOUT
New York's Subway System

FOR MORE THAN ONE HUNDRED YEARS, it has inspired jokes, legends, chance encounters, graffiti—and more than a few obscenities. On October 27, 1904, the first official New York City subway system opened. It wasn't the first in the world or in America (Boston wins that one). But New York's underground—love it or hate it—is certainly the most famous system. So watch the closing doors and test your underground knowledge in this quick quiz.

TRUE OR FALSE?

1. The subway token was introduced on the first day.

2. In 1980, New York's first subway strike shut down public transit for a week.

3. New York baseball teams have met in a "Subway Series" thirteen times.

4. According to the famous song, the "A Train" goes to Yankee Stadium.

5. The longest subway ride without changing trains is thirty-one miles.

6. If laid end to end, the subway would reach to Chicago.

7. Grand Central Terminal is the system's busiest station.

ANSWERS

1. False. The first tokens were used in July 1953. At first, riders bought tickets. Then coin turnstiles were used. When the fare went to 15 cents, turnstiles could not handle two coins, so tokens were used.

2. False. The first strike was in 1966 and lasted for twelve days. The 1980 strike lasted eleven days and is credited with introducing the practice of wearing sneakers to the office.

3. True, even though the first two meetings were not true subway series because both teams shared the old Polo Grounds.

4. False. The A train goes to Harlem. The signature Duke Ellington song was composed by Billy Strayhorn.

5. True. The famed A train will take you from Upper Manhattan to Far Rockaway, Queens.

6. True. The system has 842 track miles—only 656 of those are for passengers; the rest are for yards and shops.

7. False. Times Square handles the most passengers.

DON'T KNOW MUCH ABOUT
the U.S. Highway System

It is a familiar, even comforting sign: the white shield with six points and a black number: the U.S. Highway system. The system was born in 1925 to create an orderly system of intercity roads to replace the haphazard, sometimes privately owned roads crisscrossing America. These highways moved people around America, gave work to millions during the Depression, helped move men and munitions during World War II, and, of course, transformed the family vacation with roadside motels and diners. The U.S. Highway system became part of American lore, even surviving the Interstate system begun in the late 1950s. "Are we there yet?"

1. Which one starts in Maine and ends in Key West, Florida?

2. What is the "Old National Road"?

3. What was the old Lincoln Highway?

4. Known as the "Mother Road," which highway connected Chicago to Los Angeles?

5. In which directions do odd and even numbered highways go?

ANSWERS

1. U.S. 1.

2. Begun during the presidency of George Washington, it was the first road planned to move people west. It became U.S. 40, which eventually went from Baltimore to St. Louis.

3. The first transcontinental highway, it connected New York City to San Francisco and was replaced by U.S. 30. The Lincoln Highway was part of a network of roads, some named for presidents, which were marked with colored bands on telephone poles.

4. The fabled Route 66.

5. Odd numbers usually run north-south; even numbers run east-west and both get higher as they move west. On the modern interstate system, numbers are lower as they move west.

DON'T KNOW MUCH ABOUT
Digital Photography

CAMERAS IN OUR PHONES! Once the fancy of science fiction writers, they are now commonplace, the result of the digital camera revolution. We've come a long way since Dad once snapped a few photos of summer vacation and we'd wait for the film to come back to see if everyone's head was in the picture! Born out of videotape technology developed for television in the late 1950s, digital cameras now outsell traditional cameras around the world. Think "pixels" are creatures in fairy tales? Focus in on this quick quiz.

1. Digital or film: What's the difference?

2. What government agency led the digital revolution?

3. What is a pixel?

4. What company was first to market a digital camera?

ANSWERS

1. A digital camera, as opposed to a film that uses a chemical process, has an electronic sensor to transform still images or video into electronic data, which can be stored, transmitted, and reproduced.

2. During the 1960s, NASA converted to digital signals to map the surface of the moon. NASA also used computers to enhance the images that the space probes were sending. At the same time, digital imaging was also being put to use in spy satellites.

3. A pixel is short for "picture element," and is one of the many tiny dots that make up the representation of a picture in a computer's memory. Usually the dots are so small and so numerous that, when printed on paper or displayed on a computer monitor, they appear to merge into a smooth image.

4. The first digital camera for the consumer-level market that worked with a home computer was Apple's QuickTake 100 in 1994. Texas Instruments had patented a filmless electronic camera in 1972, the first to do so. And in 1981, Sony released the Mavica, the first commercial electronic camera. The early Mavica cannot be considered a true digital camera since it was a video camera that took video freeze-frames.

DON'T KNOW MUCH ABOUT
McDonald's

BY NOW THE BILLIONS served are nearly uncountable. But, in 1955, it all started with the first McDonald's franchise store opening in Des Plaines, Illinois, on April 15. For better or worse, the Golden Arches of McDonald's have reshaped America (and its waistline). In the process, the company became an international icon of all things American. That original store now exists only as a replica housing the McDonald's Museum. Do you want fries with this quick quiz?

1. The Des Plaines store was not the first McDonald's. Where was the true first?

2. How many shares of McDonald's would you own if you bought 100 in 1965?

3. Who portrayed Ronald McDonald in his 1963 television debut?

4. What epic product was introduced in 1968?

5. What innovation came along in 1973?

6. What is Hamburger University?

7. Who sang the McDonald's jingle in those 1970s ads?

ANSWERS

1. A hamburger stand owned by the McDonald brothers in San Bernardino, California. Milk shake machine salesman Ray Kroc saw the possibilities and the rest is history.

2. One hundred shares purchased when McDonald's went public that year would have multiplied to 74,360 shares by 2003, according to the company.

3. *Today* weatherman Willard Scott.

4. The "Big Mac," brainchild of an early franchisee, went national that year.

5. The Egg McMuffin.

6. Founded in 1961, Hamburger University is in Oakbrook, Illinois, and trains people in various aspects of the McDonald's business.

7. Barry Manilow.

DON'T KNOW MUCH ABOUT
Barbecue

OKAY, LET'S GET this straight right away. Summer is coming and the family will gather in the yard for a "barbecue." But when you cook your burgers or dogs on a grill, it is not barbecue. What a lot of people call "barbecuing" is really "grilling"—cooking over high, direct heat. To purists, barbecue specifically means to slow-cook tough cuts of meat over wood or coals (not gas!), usually with smoke. And, of course, there is the fundamental requirement to baste the meat as it cooks with a spicy sauce—also known as BBQ sauce. Grill yourself in this quick quiz.

TRUE OR FALSE?

1. The word "barbecue" comes from the French *barbe a queue* for "whiskers to tail."

2. Barbecued beef was invented by cowboys in the late 1800s.

3. Henry Ford and Thomas Edison invented the charcoal briquette.

4. The Barbecue Capital of the World is Lexington, North Carolina.

5. The colonists in Jamestown, Virginia, did barbecue.

ANSWERS

1. Probably false. While some experts like the French explanation, the word is widely thought to come from *barbacoa*, a word adapted by the Spanish from the Taino tribe. Natives of Haiti who smoked meat on a framework of sticks, the Taino also provided the word *batata* which became potato.

2. True. In one popular version, the men on cattle drives had to be fed, but only got the cheap cuts. Those cowpokes learned that leaving the meat to cook for a few hours did the trick.

3. True. Ford created the briquette from wood scraps and sawdust from his factory with the help of Edison.

4. You decide. Lexington says so. But so does Llano and Lockhart, Texas, as well as Santa Maria, California, as well as many other claimants to the title. The World Series of Barbecue takes place each October in Kansas City, which claims to have more Bar-B-Q restaurants per capita than any other city.

5. True. Adapting an Indian method, the Jamestown colonists used a pit barbecue method for cooking pork.

DON'T KNOW MUCH ABOUT
Paper Money

Can you give a buck a birthday bash? If so, light the candles! March 10 marks the birthday of America's paper currency. It was on that date in 1862 that paper money as we know it was created to help pay the costs of the Civil War when the U.S. government issued about $430 million in paper money. But the paper money trail goes way back. While coins date to the 600s B.C. in Lydia (now western Turkey), paper money began in China, probably around A.D. 600. In the 1200s, Marco Polo was amazed to see the Chinese using paper money instead of coins. What else do you know about your cash?

1. What did people call America's first paper money?

2. Who first put "In God We Trust" on American money? Which president wanted the words removed?

3. Where does "dollar" come from?

4. What is the source of the word "buck" for a dollar?

5. Who are the three non-presidents on American paper currency?

ANSWERS

1. Union soldiers called them "greenbacks," for the obvious reason that the backs of the bills were green, and the name stuck. Later paper money included "goldbacks" and "yellowbacks."

2. Treasury Secretary Salmon P. Chase who created the greenback in 1862. Teddy Roosevelt later wanted the words removed. He thought they were sacrilegious and unconstitutional, but the use of the words has been consistently upheld by various courts over the years.

3. Our word comes from the German word *thaler*, a shortened form of Joachimsthaler, named for the area in Bohemia where coins were once minted.

4. The American slang for dollar probably comes from the skins of animals that were labeled "bucks" and "does." Larger more valuable skins were "bucks." But some suggest it comes from a marker in a poker game which designates whose turn it is to deal—hence "pass the buck" means to turn responsibility over to someone else.

5. Alexander Hamilton on the $10 note; Benjamin Franklin on the $100 note; and Salmon P. Chase on the $10,000 note, which is no longer printed or in circulation.

DON'T KNOW MUCH ABOUT
the Phonograph

WHAT BECAME OF all your old "45s" and albums? Gathering dust, no doubt, unless you sold them at a tag sale or tossed them out after the compact disc arrived in the 1980s. But it was on February 19, 1878, that Thomas Edison (1847–1931) received a patent for the cylinder phonograph. With it, the genius innovator launched another industry, along with electric lights and motion pictures, to mention just two others, that changed America. What do you know about that quaint but obsolete antique, the phonograph? Spin this quick quiz.

1. What was Edison's first recording?

2. What nickname did the phonograph earn for Edison?

3. How do phonographs make sound?

4. How did phonograph records become known as "albums"?

5. Where did the nickname "wax" for records originate?

Answers

1. While doing research on the telephone in 1877, Edison shouted the verses to "Mary Had a Little Lamb" into a hand-cranked speaking machine. He was astonished when the machine, using grooved metal cylinders, played his voice back.

2. The nearly miraculous device made Edison world famous as the "Wizard of Menlo Park."

3. In a process called analog disc recording, an analog (likeness) of the original sound waves was stored as jagged waves in a spiral groove on the surface of a plastic disc. As the disc turned, a needle riding along the groove vibrated. These vibrations were transformed into electric signals and converted back into sound by speakers.

4. The short playing time of 78 RPM records meant that a long symphony or opera required as many as five discs, packaged in brown paper sleeves inside a leather-bound book resembling a photo album. Gradually, people referred to all long-playing (LP) recordings as "albums."

5. In 1885, two scientists improved upon Edison's invention by recording on cardboard cylinders coated with wax which produced better sound. The use of shellac and later plastic further improved sound and durability.

DON'T KNOW MUCH ABOUT
Microwave Ovens

It started with melted chocolate. The name Percy Spencer may not be as familiar as Edison, but his accidental discovery of a melted candy bar in his pocket revolutionized the kitchen and launched millions of bags of popcorn. In 1946, Spencer was working with a machine developed during World War II to improve British radar. After he found the melted candy, Spencer, a high school dropout and self-taught engineer, realized that waves of energy could heat food and he experimented with popcorn and eggs. According to one industry group, there are now microwave ovens in more than 90 million American homes. Can't live without your microwave? Nuke this quiz.

TRUE OR FALSE?

1. Microwave ovens "nuke" food with radio waves.

2. Microwave ovens "leak" radiation.

3. The first commercial microwave was named the "Radarange."

ANSWERS

1. True. Microwave ovens use a form of electromagnetic energy, all of which can be characterized as waves, including radio waves and visible light waves. Microwaves are the shortest of radio waves. Microwaves safely produce heat but food does not become radioactive.

2. False. A properly built and maintained microwave oven leaks so little microwave power that most scientists believe you needn't worry about it. There are inexpensive leakage testers available that you can use at home for a basic check. Or you can take your microwave oven to a service shop and have it checked with an FDA certified meter. It's only if you have dropped the oven or injured its door in some way that you might have cause to worry about standing near it.

3. True. In 1947 Raytheon demonstrated the first microwave; the name was the result of an employee contest. Designed for restaurants, it was large and expensive—more than $5,000 at the time. Domestic versions were introduced in 1967, but did not catch on until the 1970s.

DON'T KNOW MUCH ABOUT
the Atomic Bomb

ATOMIC WEAPONS have only been used twice in human history. On August 6, 1945 the United States exploded the first atomic bomb over Hiroshima, Japan, killing more than 130,000 people. On August 9, a second American atomic strike hit the Japanese port city of Nagasaki. On August 14, Japan surrendered unconditionally, ending the war. Since then, the debate over the use of these awesome weapons to defeat the Japanese without a costly invasion has continued. What do you know about the development of America's atomic weapons? Take this quick quiz.

1. What physicist's letter to President Roosevelt set in motion America's wartime program to develop an atomic bomb?

2. Though research was carried out all over the country, what was the name of the secret atomic weapons program?

3. Who was the director of the top secret atomic program?

4. The untested, uranium bomb dropped over Hiroshima was nicknamed "Little Boy" and was dropped by what aircraft?

5. What was different about the bomb dropped over Nagasaki?

ANSWERS

1. Albert Einstein, writing in 1939 at the suggestion of fellow European-emigré scientist Leo Szilard. Many of the key figures in the U.S. atomic program were European exiles who had escaped the Nazi regime.

2. The Manhattan Project was the largest single enterprise in the history of science. Work was done in nineteen different states and Canada. Other key sites included Chicago, and the newly created "atomic cities" Oak Ridge, Tennessee, Los Alamos, New Mexico, and Hanford, Washington.

3. General Leslie Groves of the Army Corps of Engineers. He code-named the Manhattan Project. The scientific director was American physicist J. Robert Oppenheimer, who later opposed development of a hydrogen bomb on moral and technical grounds and was suspended from the Atomic Energy Commission as a security risk.

4. The Enola Gay.

5. It was a plutonium-based atomic bomb.

DON'T KNOW MUCH ABOUT
the Automobile

SAY "AUTOMOBILE" and most people think of Henry Ford as the auto's inventor. But the motorcar has a long and decidedly un-American past. The first motorcars date to the late nineteenth century in Europe and America soon followed. But one landmark in America's love affair with the car came in November 1899, when the first U.S. National Automobile Show opened in New York's Madison Square Garden. History does not record whether the tradition of models in evening gowns showing off the new cars got its start back then. What do you know about the auto?

TRUE OR FALSE?

1. The world's first successful gas-driven motor vehicle is credited to Germany's Karl-Friedrich Benz in 1885.

2. America's first motorcar was the Oldsmobile "Runabout."

3. Introduced in 1908, Ford's Model T "came in any color you wanted, as long as it was black."

4. In 1914, Ford made headlines by paying his workers $5 per day.

ANSWERS

1. True. Benz developed a three-wheel motorcar that reached a speed of nine miles per hour and he began to advertise his "motor carriages" in 1888. Benz later merged with another German auto pioneer, Gottlieb Daimler, who introduced his "Mercedes" in 1901; it was named after the daughter of one of Daimler's investors.

2. False. Credit goes to the Duryea Brothers of Springfield, Massachusetts, who produced the first American internal combustion auto in 1892. They offered the first motorcar for public sale in 1896. Ransom Olds began production of his curved dash autos in 1901.

3. True. The all-black Model T, which sold for $850.50, went on to become one of the best-selling cars of all time, with 16,536,075 produced between 1908 and 1927.

4. True. By introducing the moving assembly line, Ford was able to produce an auto every twenty-four seconds and cut prices on his cars to under $300. He could well afford to give his workers a raise. At the time, Ford was personally making about $25,000 per day. He also knew that his workers would then be able to go out and buy one of those cars.

SPACE AND
THE NATURAL
WORLD

DON'T KNOW MUCH ABOUT
Pluto

UNTIL THE 2006 DECISION by the International Astronomical Union to reclassify some celestial objects as "dwarf planets," Pluto was among a very exclusive nine-member club. Before Pluto's demotion, schoolchildren could learn "My Very Excellent Mother Just Served Us Nine Pizzas." That's one way to remember the names of the planets in order moving out from the sun. In that list, the "P" in "Pizza" stands for Pluto, the last of the nine planets to be discovered. Pluto was found in 1930, when astronomer Clyde Tombaugh picked out its image in photographs. What else do you know about this "dwarf" planet, which is an average of more than 2.76 billion miles from the sun? Take this quick quiz.

1. Is Pluto really a planet?

2. Is Pluto always farthest from the Sun?

3. Is Pluto named for Mickey Mouse's dog?

4. Does Pluto have moons?

5. How long does it take Pluto to revolve around the sun?

6. Besides Pluto, what are the other eight planets in their correct order?

Answers

1. Not according to a group of astronomers at the International Astronomical Union, the group charged with naming space discoveries. In their definition of a planet, the IAU classified Pluto and two other objects in the solar system as "dwarf planets." But many astronomers disagreed with the new definitions and categories so Pluto may one day be back on the list.

2. No, sometimes its orbit falls inside the orbit of Neptune—usually the second most distant planet from the sun—and it is actually closer to the sun than Neptune is. This happened most recently between 1979 and 1999.

3. No. Pluto is named for the Roman god of the underworld.

4. Three. Charon is named after the mythical boatman who ferried the dead to the underworld. It has a diameter of about 740 miles. Two smaller moons, Nix and Hydra, were discovered in 2005.

5. Nearly 248 Earth years.

6. Mercury, Venus, Earth, Mars, Jupiter, Saturn, Uranus, and Neptune.

DON'T KNOW MUCH ABOUT
the Moon

ON JULY 20, 1969, Apollo 11 astronaut Neil Armstrong became the first to do what people had dreamed about since the dawn of civilization—to step on the Moon. For centuries before that, the Moon was the most inspirational heavenly body, firing the human imagination in myth and religion as well as inspiring some silliness, such as the idea of a "Man in the Moon." Still think it's made of green cheese? Launch into this quiz.

1. How many Americans have walked on the Moon?

2. Where did the Moon come from?

3. What is a lunatic?

4. Is there a connection between the full moon and accidents, crimes, and delivery room activity?

5. Do full moons produce werewolves?

Answers

1. Ten astronauts on five separate Apollo flights actually landed and explored the Moon.

2. There are two best guesses. One theory holds that Earth's gravity captured a small planet during the earliest period of the solar system's formation. Most scientists believe the Moon is actually formed from a piece of Earth, since they are approximately the same age. About 4.45 billion years ago, an object the size of Mars may have slammed into the recently formed Earth, blasting out fragments of the young planet in the form of vaporized hot rock. These fragments came together to form the Moon.

3. The word *lunatic*, literally meaning a "moonstruck person," was used in English as early as 1290, but its use goes back to Roman times when it was popularly believed that the mind is affected by the Moon.

4. Despite popular belief, many studies have shown there to be no connection between any human behavior, including the homicide rate, births of babies, or emergency room admissions.

5. Another moon myth, a werewolf is a person who is transformed, voluntarily or involuntarily, into a wolf under the influence of a full moon. A lycanthrope, often used to describe werewolves, refers to someone who suffers from a mental disease of fantasizing being a wolf, a mental disorder termed lycanthropy.

DON'T KNOW MUCH ABOUT
the Universe

IN NOVEMBER 1957, the Soviet Union astonished the world by sending the first animal into space, a little terrier named Laika, aboard Sputnik 2. It was the second of two shocks, since the Soviets had already launched the first man-made satellite, Sputnik 1, a month earlier. This pair of Soviet successes created a wave of Cold War fear and panic in America and ignited the Space Race between the rivals. Half a century later, Americans and Russians live and work together in space, an unthinkable idea back in 1957. But when it comes to understanding the universe around us, most of us are still lost in space.

TRUE OR FALSE?

1. Laika, the space pooch, was safely returned to Earth.

2. Dinosaurs were only the most recent victims of a deadly collision with space debris.

3. So far, no other planets have been found outside our solar system.

4. Mars, our nearest neighbor, is the fastest orbiting planet.

5. The Mercury 13 are a group of moons orbiting Mercury.

6. The Big Dipper and Orion's Belt are two of the eighty-eight constellations.

7. The largest planet in our solar system is Jupiter.

8. Famed astronomer Edwin Hubble was the first person to see Saturn's rings.

ANSWERS

1. False. The dog, whose name means *barker* in Russian, was put to sleep with a remote controlled injection, the first casualty of the Space Age.

2. True. The demise of the dinosaurs millions of years ago was the most recent of at least five major mass extinction events thought to have been caused by a large object from space smashing into Earth.

3. False. As of March 2007, the count is more than two hundred and will most likely go higher.

4. False. It is Mercury, the planet closest to the Sun, a mere 36 million miles away. That proximity gives Mercury the fastest orbit around the Sun—just eighty-eight Earth days—which is why ancient astronomers named it after the speedy messenger of the gods.

5. False. The Mercury 13 were America's first women astronaut trainees, but their careers were ended due to sexist policies at NASA.

6. False. Both are "asterisms," small groupings of stars within larger constellations.

7. True. Fittingly named for the Roman king of the gods, Jupiter is largest; more than thirteen hundred Earths could fit inside this gas giant.

8. False. The Italian astronomer Galileo was the first person to view Saturn through a telescope in the 1600s and called the rings "handles."

DON'T KNOW MUCH ABOUT
Women in Space

As WOMEN ROUTINELY fly on space shuttles and work aboard the International Space Station, they rarely make headlines. That wasn't always the case. Just ask America's first woman in space, Sally K. Ride. When Ride made space history in June 1983—twenty years after cosmonaut Valentina Tereshkova ventured into space in June 1963—it was the culmination of a dream deferred for many aspiring female astronauts. Sexism at NASA and the macho attitudes of male astronauts squelched the chances for an earlier generation of women who hoped to reach space. What do you know about women pioneering the way to space? Blast off with this quick quiz.

1. Who holds the American record for time spent in space?

2. In 1992, what were the notable firsts of astronauts Mae Carol Jemison and N. Jan Davis aboard the shuttle Endeavour?

3. Following her historic flights, what notable investigative commission did Sally Ride join?

4. What two distinctions does astronaut Eileen Collins hold?

ANSWERS

1. Although Shannon Lucid set the American single-flight duration record of 188 days back in 1996, she has since been overtaken. In April 2007, astronaut Michael Lopez-Alegria set a new record of 197 days aboard the International Space Station and was expected to reach 214 days before his scheduled return to Earth. But he knew the record wasn't going to last. Fellow NASA astronaut Sunita Williams was expected to overtake Lopez-Alegria in July 2007. Williams also holds the record for most spacewalks by a woman—ten walks and more than sixty-seven hours of spacewalking. The single-flight record still belongs to Russian cosmonaut Valery Polyakov, who set a record of 437.7 days in space during a mission in 1994–1995.

2. Jemison was the first African-American woman in space; Davis and husband Mark Lee became the first married couple in space.

3. Ride was a member of the group that investigated the January 1986 Challenger disaster which killed seven astronauts, including Judith Resnik and schoolteacher Christa McAuliffe.

4. Collins was the first woman to pilot a shuttle in 1995 and the first to command a shuttle in 1999.

DON'T KNOW MUCH ABOUT
the Mir Space Station

WHAT GOES UP, must come down. That rule proved out when the Russian Mir space station made a scheduled return to Earth in 2001 after a long and historic career aloft. Launched in February 1986, this orbiting space station crashed back to Earth in 2001. Originally designed to last three to five years, Mir was the workhorse of the Soviet space program for fourteen years. As Mir aged, its occupants endured problems and near disasters, including a crash with another spacecraft. But this orbiting outpost helped pave the way for future space development. Watch for falling space parts and take this quick quiz.

TRUE OR FALSE?

1. Mir is Russian for "Good Luck."

2. The first living thing born in orbit was a bird hatched on Mir.

3. Mir's occupants survived on food grown aboard the space station.

ANSWERS

1. False. Mir means "Peace."

2. True. In 1990, a quail was hatched in an onboard incubator.

3. False. Although wheat was grown and fish survived in an aquarium, food was delivered by unmanned cargo craft.

DON'T KNOW MUCH ABOUT
the Moon Landing

ON JULY 20, 1969, Apollo 11 astronaut Neil Armstrong uttered those memorable words: "That's one small step for a man, one giant leap for mankind." These days, when space shuttle missions seem to elicit a collective national yawn, it is worth remembering this monumental achievement, watched in rapt attention by the whole world. How much do you know about the first manned moon landing? Take this quick quiz.

TRUE OR FALSE?

1. Armstrong was accompanied to the moon's surface by crew mate Buzz Aldrin; orbiting the moon was a third crew member, Michael Collins.

2. Upon landing on the moon, the first words signaled back to earth were, "Houston, we've hit pay dirt."

3. The Lunar Rover, a small, golf cart-like vehicle, got stuck in a lunar crater.

4. The two astronauts spent nearly a full day on the moon.

Answers

1. True. Armstrong, Aldrin, and Collins were the the crew of Apollo 11.

2. False. The touchdown message was, "Houston, Tranquility Base here. The Eagle has landed."

3. False. The Lunar Rover was first used by the Apollo 15 team in 1971.

4. True. The first lunar stay was twenty-one hours and thirty-six minutes, and the entire Apollo 11 mission lasted a little more than eight days.

DON'T KNOW MUCH ABOUT
Asteroids

IT WAS A VALENTINE sent from 160 million miles away. On February 14, 1999, NASA's Near Earth Asteroid Rendezvous (NEAR) spacecraft began sending back pictures as it orbited an asteroid named Eros. This close encounter with the twenty-one-mile-long, potato-shaped chunk of rock was part of a five-year mission aimed at better understanding asteroids. Remember, an asteroid smashing into Earth is widely thought to be responsible for the extinction of the dinosaurs. Are you lost in space when it comes to meteors, comets, and other heavenly flybys? For those who don't know their asteroid from an extraterrestrial, try this quick quiz.

1. What is an asteroid?

2. What is a comet?

3. Meteoroid, meteor, and meteorite. What's the difference?

4. Might another asteroid hit the Earth?

ANSWERS

1. Small, rocky objects, asteroids are literally minor planets orbiting the sun. More than five thousand asteroids have been discovered in our solar system and scientists believe there may be twice that many in the solar system.

2. Comets are small objects of ice, rock, and gases that orbit the sun. The most famous comet is Halley's, which returns to visit Earth every seventy-six years.

3. Meteoroids are small solid objects traveling through space. When they enter the Earth's atmosphere, they are called meteors (also known as shooting stars). Most of these burn up, but if a meteor makes it all the way to the surface of Earth, it's called a meteorite.

4. Yes, but the question is when. Scientists who track Near-Earth asteroids say a collision won't likely occur within the next few thousand years.

DON'T KNOW MUCH ABOUT

Volcanoes

TALK ABOUT A BIG BANG! On August 26, 1883, the Indonesian volcanic island of Krakatau (or Krakatoa) erupted in one of the greatest explosions in recorded history. The blast was heard more than three thousand miles away, and the volcano generated large waves, or tsunamis, that killed more than thirty-six thousand people. The blast threw up a cloud of dust that obscured the sun and global temperatures dropped below normal for five years. What else do you know about nature's most destructive force, named for Vulcan, the Roman god of fire?

TRUE OR FALSE?

1. The largest volcano on earth is Mexico City's Popocatepetl.

2. Though colossal, the Krakatoa eruption was not the largest ever.

3. About five thousand volcanoes are known to have erupted on land around the world.

4. Five hundred million people worldwide are threatened by volcanic activity.

5. Krakatoa was the deadliest volcano in history.

6. An American volcano has been erupting continuously for more than twenty years.

ANSWERS

1. False. Hawaii's Mauna Loa volcano rises to more than fifty thousand feet (that's 9.5 miles!) above its base, which is under the sea. Mexico's active Popo, which occasionally sends a plume of ash over Mexico City, is 17,930 feet tall.

2. True. The explosion on the Greek isle of Santorini is believed to have been six times greater than Krakatoa. The famed eruption of Italy's Mt. Vesuvius that buried Pompeii in A.D. 79 was also larger.

3. False. There are about 540 volcanoes known to have erupted on land; however, the number beneath the sea is not known.

4. True. According to the U.S. Geological Survey, at least that many people live with the nearby threat of volcanic activity. Between 1980 and 1990, at least twenty-six thousand people were killed by volcanic activity.

5. False. Tambora in Indonesia killed an estimated ninety thousand people in 1815; however, most of them died indirectly from starvation, disease, and water contamination in the eruption's aftermath.

6. True. Hawaii's Kilauea has been erupting since January 1983.

DON'T KNOW MUCH ABOUT
the Vernal Equinox

What does it mean when the vernal equinox takes place? The term *equinox* is Latin, meaning "equal night," and means either of the two days of the year when the sun is directly above the equator. At these times, day and night are of nearly equal length everywhere on Earth. "Vernal" means "of spring." In simpler terms, spring has sprung! (At least in the Northern Hemisphere; below the Equator, autumn begins.) As the days pass, the sun will appear to move higher in the northern sky until the summer solstice in June. While that is the scientific explanation for the equinox, the arrival of spring has also inspired many of the greatest poets. In this quick quiz, match these lines about spring to the five famed poets who wrote them: William Shakespeare, Percy Shelley, Emily Dickinson, Alfred Tennyson, and T. S. Eliot.

1. "A little Madness in the Spring is wholesome even for the King."

2. "When birds do sing, hey ding a ding, ding; Sweet lovers love the spring."

3. "If winter comes, can spring be far behind?"

4. "April is the cruelest month, breeding lilacs out of the dead land . . . stirring dull roots with spring rain."

5. "In the spring a young man's fancy lightly turns to thoughts of love."

ANSWERS

1. Emily Dickinson.

2. William Shakespeare.

3. Percy Shelley.

4. T. S. Eliot.

5. Alfred Tennyson.

DON'T KNOW MUCH ABOUT
Earthquakes

ON AUGUST 17, 1999, western Turkey was shaken by one of the most deadly earthquakes in modern times. It was a horrific reminder that such quakes are often unexpected, can sometimes change history, and, of course, are potentially very deadly. Over the course of the past four thousand years, earthquakes have killed more than 13 million people. The Chinese have been recording them since around 110 B.C. But what do you know about the earth moving beneath our feet? Hold on tight with this quick quiz.

TRUE OR FALSE?

1. All earthquakes occur within the area known as the "Ring of Fire."

2. History's most deadly earthquake occurred in China.

3. An earthquake once changed the course of the Mississippi River.

4. America's deadliest earthquake was Alaska's 1964 "Good Friday" tremor.

5. Each year, there are literally millions of earthquakes around the world.

6. The most widely used measurement of earthquakes is the Beaufort scale.

ANSWERS

1. False. While many earthquakes occur along the Ring of Fire, a line of heavy volcanic activity circling the Pacific Ocean, earthquakes can occur all over the world.

2. True. In 1556, some 830,000 people died in Shaanxi, China; in 1976, another 255,000 died in Tangshan, China, the worst modern quake death toll.

3. True. One of the most powerful earthquakes in American history occurred between 1811 and 1812. Centered in New Madrid, Missouri, the quake sent large waves along the Mississippi and altered the river's course.

4. False. Although it was the most powerful in America, it was not the deadliest. That distinction belongs to San Francisco's great quake of 1906, which killed 503 people and started fires that killed hundreds more.

5. True, according to the U.S. Geological Survey; but most are small and occur undersea or away from populated areas and go unreported.

6. False. The Richter scale, devised in the 1940s, measures the energy released in a quake. The Beaufort scale measures wind velocity.

SPORTS

DON'T KNOW MUCH ABOUT
DON'T KNOW MUCH ABOUT
NASCAR

SO YOU THINK a valance has something to do with drapes? That marbles are a kid's game? You obviously haven't been hanging around a NASCAR track. The National Association for Stock Car Automobile Racing was formed in 1948 to race the kind of cars that people actually drove. (The type of cars driven at the Indy 500 are called Formula One racers.) The events run by NASCAR have grown into America's largest spectator sport. Traditionally based in the Southeast, NASCAR has spread across the map. What do you know about this American pastime? Start your engines and buckle up for a quick quiz.

1. **Where did NASCAR get its start?**

2. **What is the northernmost NASCAR raceway?**

3. **What is the smallest state with a NASCAR track?**

4. **Where is the largest sports facility in the Southeast?**

5. **Which track is located near the Chisholm Trail, the famed route of America's legendary cattle drives?**

6. **So what are "marbles" and "valances?"**

ANSWERS

1. On the famed Daytona beach and road course in the late 1940s. Since 1959, Daytona International Speedway has hosted the Daytona 500, one of NASCAR's premier events.

2. New Hampshire International Speedway in Loudon, New Hampshire. The two NASCAR events run there are New England's largest spectator sporting events, with 101,000 in attendance at each race, which equals about 10 percent of the state's population.

3. Delaware's Dover Downs International Speedway, home of "The Monster Mile." With seating for more than 140,000, it could hold nearly 20 percent of Delaware's 783,00 people.

4. Lowe's Motor Speedway in Concord, North Carolina, has 167,000 permanent seats and capacity for nearly fifty thousand more spectators in the infield area. It was also the first sports facility in America to offer year-round living accommodations when it offered forty condominiums for sale above Turn One in 1984.

5. Texas Motor Speedway in Fort Worth is a NASCAR haven with seating for 150,061 fans along the historic Chisholm Trail.

6. "Marbles" are bits of rubber shaved off tires and dirt and gravel blown around the track by passing vehicles. A "valance," or "front air dam," is a panel that extends below the front bumper.

DON'T KNOW MUCH ABOUT
the NFL

OH, FOR THE GOOD old days of the Canton Bulldogs, Dayton Triangles, and Decatur Staleys (named for their owner). These were a few of the professional football teams that met in Canton, Ohio, on September 17, 1920, and formed the American Professional Football Association. In 1922, the organization was renamed the National Football League (NFL). More than eighty-five years of passes, fumbles, and instant replays have made football America's favorite game. The Staleys? They moved to Chicago and became the Bears. What else do you know about the NFL? Tackle this quick quiz.

1. Earl "Curly" Lambeau's boss gave him $500 to help equip a team. What was this early "corporate sponsor"?

2. What was the distinction of the Oorang Indians?

3. Who was the first president of the association later known as the NFL?

4. When was the NFL integrated?

5. What turning point in pro football was reached in 1970?

6. In 1948, a player added a handmade touch to his helmet. What was it?

ANSWERS

1. The Indian Packing Company for whom the Green Bay Packers were named. They were later owned by the Acme Packing Company.

2. This early team, from Marion, Ohio, was made up of all American Indians, and was named for its sponsor, the Oorang dog kennels.

3. Jim Thorpe, the most famous athlete of his day, and a player on the Oorang Indians.

4. Black players, including the famed actor-singer Paul Robeson, played in the league's early years. But the NFL was completely segregated by 1934. The first modern black professional football players were Marion Motley and Bill Willis who broke the sport's color line in 1946 with the Cleveland Browns of the rival All-America Football Conference. The first blacks to sign with the NFL were Woody Strode and Kenny Washington of the Los Angeles Rams, also in 1946.

5. The rival National Football League and American Football League merged into what is now the NFL, with its two conferences, National and American.

6. Rams player Fred Gehrke painted horns of the Rams' helmets, the first helmet emblem.

DON'T KNOW MUCH ABOUT
Babe Ruth

SEPTEMBER WAS A good month for Babe Ruth. On September 5, 1914—batting for the minor League Providence Grays—baseball's greatest player hit his first professional home run. Next season he would be with the Red Sox— though not for long, as Boston fans know well—and make American sports history in a storied career. Homer Number 100 came in September 1920 and Number 400 on September 2, 1927. Back in the Roaring Twenties, "The Bambino" almost single-handedly saved baseball after the World Series scandal of 1919. What do you know about "the Sultan of Swat?" Step up to the plate for this quick quiz.

TRUE OR FALSE?

1. Ruth was a Boston native.

2. Ruth was a notorious child, often in trouble.

3. He was named "Babe" after the famed Ox of Paul Bunyan fame.

4. Ruth was sold to the Yankees to finance a Broadway show.

5. Ruth finished his career with his hometown Orioles.

ANSWERS

1. False. George Herman Ruth was born in Baltimore in 1895. His parents ran a saloon near the Camden Yards railway station, now site of the Orioles home field stadium. It is also the site of the Babe Ruth Museum.

2. True. He skipped school, ran the streets, and committed petty crime. By age seven, he was drinking and chewing tobacco. His parents sent him to St. Mary's Industrial School for Boys, where a priest, Brother Matthias, taught him baseball and was the major influence on his life.

3. False. In 1914, Jack Dunn signed nineteen-year-old Ruth to pitch for his minor league Orioles and took him to spring training, where he was somewhat derisively nicknamed "Dunn's Babe."

4. Partly true. The failure of the Red Sox to make the 1919 World Series, a large payroll, along with Red Sox owner Harry Frazee's failings as a theater promoter meant that he needed cash to stay afloat. He offered the best of his players to the rival Yankees. For a sum of $125,000 and a loan of more than $300,000 (secured on Fenway Park itself), Ruth was sold to the Yankees on January 3, 1920. The deal became fabled as the "Curse of the Bambino," the reason the Red Sox did not win a World Series until 2004.

5. False. The Yankees released Ruth after the 1934 season, and he ended his playing career in 1935 with the Boston Braves. In the final game he started in the outfield for Boston, Ruth hit three home runs. He died in August 1948 at age fifty-three.

DON'T KNOW MUCH ABOUT
the Ancient Olympics

IT WAS THE LONGEST running show in ancient history. Every four years, for nearly twelve hundred years, hundreds of naked athletes and thousands of ancient sports fans gathered for the Olympic festival. When the Summer Games returned to Athens in 2004, it was a reminder of the glory that was Greece. But the modern games have little relation to what has been described as Woodstock, Mardi Gras, and a religious revival all rolled into one. What do you know about the ancient Olympics?

1. The ancient Olympics were held near Mt. Olympus.

2. The Marathon commemorates the famous run of a messenger delivering news of a Greek victory over the Persians.

3. The ancient games included chariot races, running, wrestling, javelin, and discus.

4. Women could not watch or participate in the ancient games.

5. The ancient games were banned by the Christian emperor of Rome in 394.

ANSWERS

1. False. Mt. Olympus, the legendary home of the gods, is in northern Greece. The site of the games, the sanctuary of Olympia, was in southern Greece.

2. False. The story of a runner sent from the Battle at Marathon to Athens—a distance of 26.3 miles—is a legend. Historians do say a messenger ran from Athens to Sparta (153 miles) to request help when the Persians landed at Marathon.

3. True. All of these events, plus boxing and long jumping, were part of the ancient games. There were no team sports or swimming events. The highlight was the four-horse chariot race, a dangerous and usually bloody event.

4. False. The Olympics were mostly all-male events, but a Spartan woman won the chariot race twice, and unmarried women were permitted to attend the games. There were also plenty of female prostitutes on hand. There are also accounts of a separate sporting event for girls, known as Hera's Games. In the modern games, women competed only in tennis in 1900. Women's track and field debuted in 1928. There was no women's marathon until 1984.

5. True. Emperor Theodosius II banned all pagan festivals, including the games, although they may have continued briefly. The Temple of Zeus at Olympia was later burned, and earthquakes and floods eventually buried the entire site.

DON'T KNOW MUCH ABOUT
the New York Yankees

IT WAS A SIMPLER time back in 1903 when baseball fans didn't have to worry about labor negotiations. That year, the American League approved a new team in New York and their inaugural game was played on April 22, 1903. Since then, the championships have been piled high by the New York Yankees, the most successful franchise in sports history. The roster of Yankee greats is a Hall of Fame in itself—Ruth, Gehrig, DiMaggio, Mantle, Mays, and Berra, to name just a few. What do you know about the most beloved—and hated—team in baseball history? Step up to the plate for this quick quiz.

TRUE OR FALSE?

1. The Yankees were originally known as the Highlanders.

2. The long string of Yankee championships began in their first season.

3. The word "Yankee" comes from the Dutch words for "John Cheese".

ANSWERS

1. True. Their first stadium, Hilltop Park, was a hastily built wooden structure on one of the highest points in Manhattan.

2. False. Their first American League pennant did not come until 1921, and the first World Series win was in 1923. Before then, rival Boston was far more successful, having won six of the first eighteen American league pennants and four World Series between 1912 and 1918.

3. True. It was a mild insult used by the Dutch for English colonial settlers.

DON'T KNOW MUCH ABOUT
the World Series

IT HAS SURVIVED world wars and earthquakes. They call it the World Series even though it is almost exclusively American. And while the Fall Classic is still played in October, it is getting perilously close to November. (Somehow calling Reggie Jackson "Mr. November" just doesn't have the right ring.) An integral part of Americana for more than a century, the World Series still captivates the nation. What do you know about the championship of baseball? Take this quick quiz.

1. In what year was the first World Series played?

2. Since that first World Series, how many times has it been canceled?

3. "Say it ain't so Joe" were the famous words supposedly uttered to "Shoeless" Joe Jackson, one of eight Chicago White Sox players charged with throwing the World Series for money. In what year was the infamous "Black Sox" scandal?

4. Everyone knows the Yankees have won the most World Series championships. What team is Number Two?

Answers

1. In 1903, the Boston Pilgrims (Red Sox) of the upstart American League defeated the National League's Pittsburgh team.

2. Twice. In 1904, prior to an agreement to play the series every year, the New York Giants refused to play against Boston; in 1994, a baseball labor dispute forced cancellation.

3. 1919. The victory of Cincinnati over Chicago was declared invalid. Although a court later found the Chicago players not guilty, all eight players were still barred from baseball.

4. The St. Louis Cardinals, with ten championships as of 2006, follow the Yanks with 26. (The Yankees have also lost the most world series–13!)

Sex and Sports

BEFORE THERE WAS Mia Hamm dueling Michael Jordan in TV commercials. Before there was professional women's basketball. Before women hockey players laced up skates in the Olympics, there was Billie Jean King. On September 20, 1973, King helped break down one more barrier for women when she accepted a challenge to play tennis against a man. Egged on by showman and former champion Bobby Riggs, King agreed to a nationally televised match against the aging veteran. Following a huge publicity buildup, the world watched this "Battle of the Sexes," as Billie Jean King won handily, in a match that proved little about the superiority of either sex. But the moment was a turning point in the recognition of female athletes, making the likes of the WNBA, the women's World Cup, and female hockey accepted facts. What else do you know about other great women in sport?

1. Who holds the most tennis tournament titles, male or female?

2. An Olympic track and field star and champion golfer of the 1930s through the 1950s, who is often called the greatest woman athlete?

3. What distinction does Anne Meyers hold?

4. Who was the first black woman to win a Wimbledon title?

5. Who holds the most New York City Marathon titles, male or female?

Answers

1. Martina Navratilova with 341 singles and doubles titles, not counting mixed doubles.

2. Babe Didrikson Zaharias, who won two gold medals in the 1932 Olympic Games and then dominated women's golf for the next three decades.

3. She was the first woman to go to a National Basketball Association training camp, trying out with the Indiana Pacers.

4. Althea Gibson in 1957 and again in 1958.

5. Norway's Grete Waitz won the race nine times between 1978 and 1988.

DON'T KNOW MUCH ABOUT
Basketball

THE WORLD CHAMPION is crowned in June and the game is played in the Summer Olympics. But basketball was conceived as a winter sport. The roots of "hoops" go back to December 1891, when James A. Naismith devised the game in Springfield, Massachusetts, now home of the Basketball Hall of Fame. Trying to create a vigorous, but theoretically non-contact, indoor game, Naismith hung the legendary half-bushel peach baskets in a YMCA gymnasium where he was a physical education instructor. William R. Chase broke open a scoreless game when he connected on the first basket in history, leading his nine-man team to a 1–0 victory. Word of the sport spread quickly, and basketball flourished in local gyms, college campuses, and professional leagues. By the way, the inventor of this All-American sport was a Canadian. Think you know your roundball? Try this quick quiz.

1. Why are basketball players called "cagers"?

2. Who played the first five-man college game?

3. What type of ball was first used to play basketball?

4. Who was the first black player in the NBA?

5. Who is the NBA's career scoring leader?

6. Besides conceiving basketball, what other sporting device did Naismith create?

ANSWERS

1. In the early years, courts were surrounded by chicken-wire cages to protect players from unruly fans.

2. The University of Chicago defeated the University of Iowa in the first five-man per side game in 1896. Before that, teams used as many as nine players.

3. The first games were often played with soccer balls.

4. Charles Cooper who joined the Boston Celtics in 1950.

5. Kareem Abdul-Jabbar with 38,387 points. Abdul-Jabbar is second in games played behind Robert Parish, and third in rebounds behind Wilt Chamberlain and Bill Russell.

6. Naismith is credited with inventing the football helmet.

DON'T KNOW MUCH ABOUT
Jackie Robinson

In the "Show Me the Money" era of million-dollar athletes such as Michael Jordan and Tiger Woods, it is easy to forget that only half a century ago there was no place for the black athlete on the playing fields of America. Jack Roosevelt Robinson changed that forever when he became the first African American to play major league baseball. Born on January 31, 1919, Jackie Robinson didn't invent anything. He didn't give memorable speeches. He led no protest marches. But Robinson, who died in 1972, helped change the face of the country by displaying courage, dignity, and remarkable skills in the face of vicious racism. How much do you know about Jackie Robinson?

1. In what year did Robinson break major league baseball's "color line"?

2. With what team did he spend his entire major league career?

3. In 1949, Robinson won the National League batting title and what other honor?

4. What position did Jackie Robinson play?

5. A four-sport college athlete, what other professional sport did Robinson play?

ANSWERS

1. Robinson played his first major league game on April 11, 1947.

2. The Brooklyn Dodgers from 1947 to 1956.

3. He was named the National League's Most Valuable Player.

4. Robinson was an outstanding second baseman.

5. Before his baseball career, Robinson played football in the Pacific Coast League.

DON'T KNOW MUCH ABOUT
Muhammad Ali

ONE OF THE MOST recognized people on the planet, he could "float like a butterfly/sting like a bee." Politicians now flock to be photographed at his side. But it was a different story in 1964 when Cassius Clay knocked out Sonny Liston on February 15 to become heavyweight champion of the world. In a day when athletes, especially African-American ones, were seen and not heard, Cassius Clay would become Muhammad Ali—and the world had never heard or seen anything quite like him. What do you know about the man called "the Greatest," diagnosed in 1984 with Parkinson's disease, who was once the world's most controversial athlete? Step into the ring for this quick quiz.

1. Where, when, and for what did he win Olympic gold?

2. Why and when did he change his name?

3. How did he lose his title in 1967?

4. How many times did Ali win the heavyweight title?

5. What notable boxer linked to Ali shares his birth date?

ANSWERS

1. As Cassius Clay, he won the light heavyweight title at the 1960 Olympics in Rome. He later said he threw the medal into the Ohio River.

2. Born Cassius Marcellus Clay in Louisville, Kentucky, he adopted the Black Muslim religion and, renouncing what he called his "slave name," changed it to Muhammad Ali.

3. In 1967, boxing groups stripped Ali of his title when he refused induction into the United States Army and was convicted on charges of refusing induction. After this conviction, Ali did not box for 3½ years. In 1971, the Supreme Court of the United States reversed Ali's conviction.

4. Ali won the heavyweight title for a fourth time when he defeated Leon Spinks. He announced his retirement in 1979, but then fought Larry Holmes for the World Boxing Council version of the title in 1980. Holmes defeated Ali by a technical knockout.

5. Joe Frazier, who fought Ali in three memorable bouts, was born on January 17, 1944, two years after Ali was born. Their first 1971 bout, called the Fight of the Century, was won by Frazier.

ENTERTAINMENT

DON'T KNOW MUCH ABOUT
Alfred Hitchcock

FIRST THERE WAS THE MUSIC ("Funeral March of a Marionette" by Gounod). Then came the distinctive, portly caricature profile. And finally, the familiar, "Good evening," spoken in Alfred Hitchcock's singular style. Long before *The X Files*, a generation of television watchers was hooked on the strange mysteries found each week on *Alfred Hitchcock Presents*, which made its television debut in October 1954. The masterful director of classic thrillers was already famous for such films as *Rebecca, The Man Who Knew Too Much,* and *The Lady Vanishes*. But he brought a macabre, irreverent sense of humor to the predictable menu of variety shows and comedies of 1950s television. Hitchcock produced and hosted the thirty-minute series from 1955 to 1962. What else do you know about the genius behind *Psycho* (see the related quiz on page 213) and *Rear Window* who was born in 1898?

1. What famous children's book author wrote several stories produced by Hitchcock, including a memorable episode in which detectives are served leg of lamb, the weapon in the murder they are investigating?

2. What future star of television and films (including *The Great Escape*) was featured as an unlucky gambler in an episode selected as one of *Entertainment Weekly* magazine's "Best 100 Television Shows"?

3. How many Academy Awards did Hitchcock win as Best Director?

4. Which 1934 Hitchcock film did he remake in 1956?

ANSWERS

1. Roald Dahl, famed for such children's classics as *James and the Giant Peach*, *Willy Wonka*, and *The Witches*.

2. Steve McQueen. Many other famous actors, including Vincent Price, William Shatner, and Charles Bronson had roles on the show.

3. None. He did receive the Academy's life achievement award in 1967.

4. *The Man Who Knew Too Much*.

DON'T KNOW MUCH ABOUT
George Gershwin

HE WAS BORN Jacob Gershowitz on September 26, 1898, and American music would never be the same. The son of Russian immigrants, George Gershwin rose from playing piano for a few dollars in New York's Tin Pan Alley to become America's most celebrated composer. Often working with his lyricist brother Ira, Gershwin bridged the world of pop music, Broadway, jazz, and the classics to create some of America's most familiar and enduring music, including "Someone to Watch Over Me," "I Got Rhythm," and "Let's Call the Whole Thing Off" ("You say potata . . .") and "Summertime." What do you know about George Gershwin?

1. Made famous by singer Al Jolson, what was Gershwin's first big hit?

2. What 1928 Gershwin composition became a 1951 Gene Kelly film?

3. What was Gershwin's most famous composition for a jazz band, piano, and symphonic orchestra?

4. Which Gershwin work, based on a novel about a poor black fishing village, opened in 1935 and closed after a little more than one hundred performances?

ANSWERS

1. "Swanee."

2. "An American in Paris."

3. "Rhapsody in Blue."

4. *Porgy and Bess.*

DON'T KNOW MUCH ABOUT
Lucille Ball

IN THE BEGINNING, there was Lucy. Long before Marlo, Mary, Rhoda, Roseanne, Ellen, and other one-name female television comics, Lucille Ball blazed a trail for women in television. Whether it was stomping grapes or packing chocolates, Lucy traveled a long, hilarious path that has never been surpassed. Born in August 1911, Lucille Ball died after surgery in 1989. What do you know about the zany redhead whose escapades still delight viewers? Tune in to this quick quiz.

TRUE OR FALSE?

1. Lucy was the first woman to be portrayed as pregnant on television.

2. Despite her success, Lucille Ball never appeared on the cover of *TV Guide*.

3. Known for her red hair, Lucy was a "Red," a registered communist who was blacklisted in the McCarthy era.

4. Lucy met her husband, Cuban bandleader Desi Arnaz, when she was a Ziegfeld Follies showgirl.

5. A successful movie actress before television, Lucy had auditioned for the role of Scarlett O'Hara.

ANSWERS

1. True. Little Ricky Ricardo and real son Desi Arnaz, Jr., were "born" on the same day: January 19, 1953. The episode was the most watched show in television history to that time.

2. False. She was actually on the first issue and appeared on more *TV Guide* covers than anyone else.

3. True. Although not a party member, Lucy had once registered as a communist. She was able to denounce the charge and it did not damage her career.

4. False. They met when they costarred in a 1940 movie *Too Many Girls.*

5. True. Lucy had made many films before shifting to television and had auditioned for *Gone With the Wind.*

DON'T KNOW MUCH ABOUT
Psycho

JUST WHEN YOU thought it was safe to go back in the shower—along comes a rerun of Alfred Hitchcock's legendary thriller *Psycho*. When it opened in June 1960, the creepy tale of theft and murder at the Bates Motel broke many of the rules of movie making and went on to be considered one of the greatest movies of all time. What do you know about one of the masterpieces of Hitchcock's career? Check into this quick quiz.

1. What daily household function was shown for the first time in a movie?

2. The original name of Janet Leigh's character was Mary Crane from Dallas. To what was it changed and why?

3. What was Norman Bates's unusual hobby?

4. What is the "blood" in the infamous shower scene?

5. *Psycho* marks what famous "last time" in Hitchcock's illustrious career?

6. Which new rule for movie theaters was created for the release of *Psycho*?

ANSWERS

1. Not a shower, but a flushing toilet.

2. She became Marion Crane from Phoenix after Hitchcock learned there was a real Mary Crane in Dallas.

3. Taxidermy.

4. Chocolate sauce, which showed up better than stage blood in the black and white film.

5. It was his last black and white film; afterward the others were in color.

6. Theater owners were told not to allow seating after the movie began, a policy enforced with security guards for publicity value. The movie helped establish the policy of set show times.

Disneyland

"HERE YOU LEAVE TODAY AND ENTER THE WORLD OF YESTERDAY, TOMORROW AND FANTASY." Those words greet visitors to one of America's favorite places—Disneyland, which opened officially on July 17, 1955. The brainchild of American genius Walt Disney, the "mother of all theme parks" welcomed an estimated 515 million guests in its first half century. What do you know about the park that a Mouse built? Clap your hands three times and try this quick quiz.

TRUE OR FALSE?

1. The televised opening ceremonies were hosted by Ronald Reagan.

2. Opening day was largely a disaster, later called "Black Sunday" by company executives.

3. No one has ever died in an accident at Disneyland.

4. In its fifty years, Disneyland has never closed.

5. The park was once "invaded" by "Yippies" who wanted to "liberate" Minnie Mouse.

Answers

1. True. A friend of Disney, Reagan was one of three hosts. The others were Art Linkletter and actor Bob Cummings.

2. True. Counterfeit tickets led to overcrowding and road congestion; the temperature hit 110 degrees, melting the blacktop in which women's high heels got stuck; and a plumber's strike meant drinking fountains were dry.

3. False. In the first fifty years, there were nine deaths. Two of them were officially ruled the result of park negligence; the others were ruled negligence on the part of the "guests." Since 2005, there have been several more deaths at Disney theme parks, including two on the Epcot Mission Space ride in Florida.

4. False. It has closed twice. First after President Kennedy's assassination in 1963 and again after the terror attacks on September 11, 2001.

5. True. In 1970, several protesters raised a Viet Cong flag on Tom Sawyer's island and filled one of the rides with marijuana smoke, forcing the park to shut down early.

DON'T KNOW MUCH ABOUT
the Tonight Show

IT DIDN'T START WITH "Heeeere's Johnny!" But on September 27, 1954, an American institution began and we have been staying up late ever since. Over that span, four different hosts, each with a different style, made *Tonight* part of the American scene. Blending comedy, talk, music, and animal acts, *Tonight* was the model for dozens of other late night talk shows. What do you know about this fifty-plus-year-old institution? Tune in this quick quiz.

1. Who was the first host of the *Tonight* show?

2. What made second host Jack Paar abruptly leave the show?

3. How long was Johnny Carson the host of *Tonight*?

4. Who got married on the *Tonight* show?

5. Before his success as a comic, what did Jay Leno do for a living?

6. What major recent political headline was made in 2003 on the *Tonight* show?

ANSWERS

1. Steve Allen who hosted the show until 1957.

2. Paar took over in 1957. When one of his comedy sketches was censored in 1960 over a reference to a "water closet," he walked off the show. Standards were different in those days. Paar returned but stirred controversy over an interview with Fidel Castro and broadcasting from Berlin when the Berlin Wall went up in 1962.

3. Thirty years, from 1962 to 1992.

4. One of the largest television audiences ever tuned in on December 17, 1969, to watch the ukulele-strumming crooner Tiny Tim marry "Miss Vicki."

5. After graduating from Emerson College with a degree in speech therapy, Leno held jobs as a mechanic and delivery man.

6. Arnold Schwarzenegger used the show to announce his candidacy for governor of California in August 2003.

DON'T KNOW MUCH ABOUT
Motown

AN ELEVENTH GRADE dropout turned featherweight boxer, Korean war veteran and failed record store owner doesn't sound like the biography of a man who changed America's musical tastes. But that is the story of Berry Gordy, who was born on November 28, 1929. Founded in 1959, his famed Motown Records was the home of the Supremes, Temptations, Jackson Five, and so many other classic performers who gave America a soundtrack it could dance to. What do you know about the record company with a sometimes stormy history called "Hitsville, USA"? Spin this quick quiz.

1. What was Motown's first number one record?

2. Which Motown secretary later sold a million copies of "Heat Wave" with her group?

3. What Motown superstar singer started as a drummer for Smokey Robinson and the Miracles?

4. Which Motown classic got new life as a famous television commercial?

5. With twelve Number One Pop Singles, which Motown group started out as a quartet called the Primettes?

6. Why was it called Motown?

ANSWERS

1. "Please Mr. Postman" by the Marvelettes.

2. Martha Reeves worked as a secretary before she and the Vandellas hit the charts. Their name came from merging Detroit's Van Dyke street with Reeves's favorite singer, Della Reese.

3. Marvin Gaye, who had his first Motown hit in 1962 and his first number one hit, "Heard It Through the Grapevine," in 1968.

4. "Heard It Through the Grapevine" introduced the singing California Raisins.

5. The Supremes. They struggled at first and Diana Ross worked in a department store cafeteria until they hit it big with "Where Did Our Love Go?" in 1964.

6. The company was in Detroit, the "Motor City," which is also known as "Motown."

DON'T KNOW MUCH ABOUT
Playboy

IN 1953, an otherwise obscure advertising man from Chicago pasted together a magazine on his kitchen table. He printed seventy thousand copies, hoping to sell at least thirty thousand of them at fifty cents an issue. On December 1, 1953, *Playboy* magazine was brought to life by Hugh Hefner; it was an immediate sellout. The magazine and its bunny logo would become American icons as the centerfolds, interviews, and cartoons heralded the changing sexual attitudes of postwar America. What do you know about this American sex symbol? Bare all in this quick quiz.

TRUE OR FALSE?

1. The first issue was dated December 1, 1953.

2. The magazine's original name was *Stag Party*.

3. The famed first centerfold of Marilyn Monroe was photographed by Hefner.

4. The famed bunny logo made its debut in the second issue.

5. A full frontal nude first appeared in 1959.

6. Most people just "read it for the articles."

ANSWERS

1. False. It carried no date because Hefner didn't know if there would be a second issue. Copies of that fifty-cent issue now sell for over $2,000.

2. True. The publisher of a hunting magazine called *Stag* forced Hefner to change the name.

3. False. Photographer Tom Kelly had taken the shot in 1949 and it had already appeared in calendars. Hefner paid $500 for the right to use the famous shot of Monroe on rumpled red velvet.

4. True. Chosen by Hefner because it was "frisky and playful," the bunny was created by art designer Art Paul for the second issue and has appeared on every issue since.

5. False. This breakthrough did not come until 1969.

6. Chances are that is false.

DON'T KNOW MUCH ABOUT
60 *Minutes*

THE FAMILIAR "tick-tick-tick" of the stopwatch started in 1968 and hasn't stopped yet. The investigative journalism show that changed television forever premiered on September 24, 1968, the brainchild of producer Don Hewitt. It was a Tuesday night, not a Sunday, and the preceding shows were the Red Skelton and Doris Day comedy-variety hours. Since then, the hard-hitting correspondents on the show have broken major stories, brought down companies, and often made news themselves by taking on such targets as the Army during Vietnam and the auto industry, airing historic interviews, and even televising an assisted suicide. What do you know about the show that launched the television newsmagazine?

1. Was the show always a success?

2. Who were the first anchors?

3. Who was the first woman correspondent?

4. What 1999 film was inspired by a 60 *Minutes* controversy?

5. How many Emmys has it won?

6. When did Andy Rooney join 60 *Minutes*?

ANSWERS

1. No. It struggled in the ratings and was shifted in the schedule seven times before settling into the 7 P.M. slot on Sunday night in December 1975. From 1977 through 1999 it finished in the Top Ten programs for twenty-two consecutive seasons. That record may be safe. The next longest Top Ten streak belongs to Lucille Ball whose shows had a ten-year run.

2. Mike Wallace and Harry Reasoner. Morley Safer was added in 1970 and Dan Rather in 1975. Ed Bradley joined the show in 1981 and served until his death in 2006.

3. Diane Sawyer who joined in 1984. She left to anchor the rival *Prime Time Live* on ABC in 1989.

4. *The Insider* starring Al Pacino as producer Lowell Bergman whose report on the tobacco industry was shelved by the network.

5. Through the 2004–2005 season, *60 Minutes* has won more Emmy Awards than any other news program, with a total of seventy-eight, including a Lifetime Achievement Emmy to producer Don Hewitt and the correspondents in 2003.

6. The lovable curmudgeon Rooney was with the show as a writer and producer since its premiere but did not begin his on-air segments until 1978.

DON'T KNOW MUCH ABOUT
The Catcher in the Rye

HOLDEN CAULFIELD on Social Security! Now there's a thought. *The Catcher in the Rye* was published in July of 1951, so the perennial seventeen-year-old, prep school drop-out has hit his golden years. As another crop of high school students heads to class, J.D. Salinger's tale of adolescent alienation goes with them. It is still one of America's most widely assigned books, as well as one of its most regularly censored books, according to the American Library Association. (According to the ALA, most books are "challenged" by parents but not actually banned, thanks to the efforts of librarians.) Still crazy after all these years, Holden has been the bane of many students, an inspiration for others. What do you know about this staple of reading lists, and its quirky reclusive author, more than half a century after the book appeared?

1. Which famous museum makes Holden happy to think about as he wanders around New York?

2. Holden disliked "phonies," which to him meant almost everyone, except for which character in the book?

3. Why is *The Catcher in the Rye* banned in certain schools and libraries?

4. How many other novels did Salinger write?

5. What book topped the 2005 list of books most often challenged by parents?

ANSWERS

1. The American Museum of Natural History.

2. His "kid sister" Phoebe.

3. Some people have objected to its use of profanity, which is by modern standards pretty mild.

4. None. *Catcher* is his only novel; his other books are collections of short stories, most of which appeared in *The New Yorker* magazine.

5. According to the ALA Office for Intellectual Freedom, *It's Perfectly Normal* by Robie Harris, for homosexuality, nudity, sex education, religious viewpoint, abortion, and being unsuited to age group; *Forever* by Judy Blume, for sexual content and offensive language; and *The Catcher in the Rye* by J. D. Salinger, for sexual content, offensive language, and being unsuited to age group. All three of these books, along with the *Harry Potter* series and *The Adventures of Huckleberry Finn*, were among the 100 Most Frequently Challenged Books for the Years 1990 to 2000.

DON'T KNOW MUCH ABOUT
Walt Disney

FEW PEOPLE HAVE had as much impact over the last one hundred years as a visionary artist as Walter Elias Disney, who was born on December 5, 1901. Disney was one of the greatest geniuses in American entertainment history, and his imagination and optimism have inspired and entertained millions of people worldwide. In 1923, Disney moved to Los Angeles and set up his first studio in the back of a real estate office. Disney struggled to pay expenses but gained his first success in 1928, when he released the first short cartoons that featured Mickey Mouse; later full-length animated films include the classics *Fantasia, Dumbo, Bambi, Cinderella, Alice in Wonderland, Peter Pan, Lady and the Tramp*, and *Sleeping Beauty*. Disney also pioneered television and then mortgaged everything to acquire the orange grove that became Disneyland in 1955. So Mouseketeers, put on your ears and try this quick quiz.

TRUE OR FALSE?

1. Disney did the drawing but not the voices of the cartoons.

2. In 1940, *Pinocchio* was Disney's first full-length feature cartoon.

3. During World War II, Disney produced propaganda films.

4. Disney once testified that Communists had tried to take over the studios.

5. After his death, Disney's remains were cryogenically frozen.

ANSWERS

1. False. Disney provided Mickey Mouse's voice, but after 1924, Disney no longer did the drawing necessary for his animated films. His genius lay in creating and directing the studio and later television productions.

2. False. In 1937, Disney issued the first full-length animated film to be produced by a studio, *Snow White and the Seven Dwarfs*.

3. True. Disney's studio created anti-Nazi cartoons as well as other educational films for the U.S. government.

4. True. In 1947, he told the House Un-American Activities Committee (HUAC) that Communists were behind a union effort to organize his writers and artists.

5. False. This is a popular "urban legend." A chain smoker, Disney died of lung cancer in 1965 and was cremated, his remains interred in Glendale, California.

DON'T KNOW MUCH ABOUT
Mickey Mouse

BORN ON NOVEMBER 18, 1928, the squeaky voiced "Mouse That Roared" has gone on to become one of America's most beloved characters. America's favorite rodent grew out of a cartoon character named Oswald Rabbit, created by Mickey's "father," Walt Disney, for Universal Pictures. When Walt Disney left Universal, Oswald was given a "mouse-over," and history was changed. Mickey has come a long way in the meantime, and he is now the cornerstone of an enormous international business, one of the world's most recognizable "celebrities." So put on your Mouseketeer ears and take this quick quiz.

1. What was Mickey's first film appearance?

2. What was Mickey's role in the 1941 feature film *Fantasia*?

3. How old is Mickey's good friend Donald Duck?

4. In what World War II battle did Mickey play a role?

5. What was the first feature-length cartoon created by Disney?

Answers

1. Although *Steamboat Willie* is widely considered Mickey's debut, he appeared first in *Plane Crazy*, a silent film followed by several others. But *Steamboat Willie* was the first shown to the public. The movie debuted at the Colony Theater in New York on November 18, 1928, and was the first cartoon with successfully synchronized sound. So that date is officially considered Mickey's birthday.

2. He played the mischievous and comically inept Sorcerer's Apprentice.

3. Donald Duck is old enough to receive Social Security. He made his debut in 1934, with a bit part in the animated *Wise Little Men*.

4. D-Day. Mickey Mouse's name was the password used by intelligence officers planning the invasion of Normandy.

5. *Snow White and the Seven Dwarfs* (1937).

DON'T KNOW MUCH ABOUT
the Beatles

THEY WERE JUST "four lads" playing music in Liverpool's club scene in the late 1950s. Then the Beatles burst onto the music world's main stage and changed rock and roll, hairstyles, and much more. Cutting across generational lines, the Beatles' music is now shared by aging boomers and their teenagers. In 1999, one of their hits was named the "Greatest Song Ever Written," according to a poll of famous songwriters, topping such classics as "Over the Rainbow." (See question #1. You may be surprised.) And in Liverpool, the hometown "lads" have been honored with Beatles Week, a festival and fan convention featuring tours and concerts. Know your "Fab Four"?

1. What Beatles tune was recently named the "Best Song Ever Written?" (Hint: It wasn't "Lucy in the Sky with Diamonds.")

2. The lads made four movies. What are they?

3. Their music inspired a fifth movie. What is that?

4. In 1966, John Lennon made international headlines when he made what comparison?

ANSWERS

1. No, not "Yesterday" or "Here, There and Everywhere." The winner was John Lennon's autobiographical "In My Life," off the 1965 album *Rubber Soul*.

2. *A Hard Day's Night* (1964), *Help!* (1965), *Magical Mystery Tour* (1968), and *Let It Be* (1970).

3. The animated *Yellow Submarine* (1968) for which they also supplied voices.

4. Lennon told an interviewer, "We're more popular than Jesus now. I don't know which will go first—rock 'n' roll or Christianity."

FOOD

DON'T KNOW MUCH ABOUT
Cornflakes

"HOW CAN YOU EAT anything that looks out of eyes?" Not much of a slogan for a breakfast cereal, but those were the sentiments of John Harvey Kellogg. A surgeon and Seventh Day Adventist, Kellogg was director of the Western Health Reform Institute, a sanitarium in Battle Creek, Michigan, who kept his patients on a strict vegetarian diet. In May of 1894, legend has it that Kellogg left one of his concoctions in the oven too long. His brother Will Keith convinced him to package the result, originally made from wheat, as flakes instead of grinding it. Kellogg's flaky cereal was imitated by more than one hundred rival Battle Creek cereal factories. John Kellogg later sold the commercial rights to Will Keith who started the Battle Creek Toasted Corn Flake Company, later the Kellogg Corp. What else do you know about this American breakfast institution? As part of a nutritious meal, taste this quick quiz.

TRUE OR FALSE?

1. Kellogg's original concoction of wheat, oatmeal, and cornmeal was known as Gorp.

2. Kellogg's thin flakes of corn cereal were first marketed as Sanitas Corn Flakes.

3. These Corn Flakes were an instant success.

4. One of Kellogg's patients, Charles W. Post, developed a coffee substitute called Postum and later created Grape Nuts.

5. Kellogg's exploits are featured in a novel and movie.

ANSWERS

1. False. In 1876, Kellogg introduced Granula but another firm was using that name, so it was changed to Granola.

2. True. The first cornflakes were sold by Kellogg's Sanitas Nut Food Company.

3. False. The first batch went rancid on grocer's shelves. Will Keith also thought to sweeten them with malt.

4. True. Post introduced Grape Nuts as a health food in 1897.

5. True. T. C. Boyle's *The Road to Wellville*, source of a film by the same title, offers a fictional version of these events.

DON'T KNOW MUCH ABOUT
Hamburgers

SUCCESS HAS A hundred fathers, the saying goes, and so it is with the hamburger. Long before any "Golden Arches" graced the land, there was Delmonico's, a venerable New York restaurant. There, in 1835, the first printed menu in America featured "Hamburg Steak," an expensive dish named after its city of origin in Germany. It took some time before that dish evolved into the "Hamburger" we know and love. Louis Lassen of New Haven, Connecticut, is said to be the first one to take broiled ground beef in 1899 (then at seven cents a pound) and serve it between two slices of toast. With billions and billions sold each day, we've come a long way. What else do you know about this all-American favorite? Take-out this quick quiz.

1. "I'll gladly pay you Tuesday for a hamburger today" is the signature line of which famous character?

2. In what year did McDonald's open for business?

3. Has there ever been a war between two countries with a McDonald's?

4. Are "French fries" really French?

5. Is there a difference between "ketchup" and "catsup"?

Answers

1. Popeye's cartoon pal Wimpy, whose full name is J. Wellington Wimpy.

2. In 1940, Richard and Maurice McDonald opened their first drive-in near Pasadena, California. In 1948, the brothers began to franchise the name and installed the first infrared French fry warming lights. In 1954, milk shake-machine salesman Ray Kroc bought the franchise rights from the brothers. Kroc completed the buyout in 1961 and acquired the McDonald's name for $14 million.

3. Until the allied assault on Yugoslavia, there had never been an armed conflict between two countries with McDonald's restaurants, according to *New York Times* columnist Thomas Friedman.

4. No. "French fries" originated in Belgium in the mid-nineteenth century, crossed the border into France, and arrived in America late in the nineteenth century.

5. No matter how you spell it, tomato ketchup or catsup is the same thing and has been part of the English language since the 1600s. It is derived from a Chinese word for "pickled fish brine." Want a little fish brine with those fries?

DON'T KNOW MUCH ABOUT
Ice Cream

"I SCREAM. YOU SCREAM. We all scream for ice cream!" Dripping with a thick fudge sauce of legend, the history of ice cream is kind of soupy. Myth has it that Marco Polo brought the secret of ice cream back from his China trip, but that story has been dismissed. However, the invention of ice cream is still credited to the Chinese as far back as 2000 B.C. What do you know about America's favorite summer treat?

TRUE OR FALSE?

1. America's favorite flavor is vanilla.

2. Ice cream was introduced to America by First Lady Dolley Madison at her husband's 1813 inaugural ball.

3. The ice-cream sundae was invented in 1887 in honor of Queen Victoria's Golden Jubilee. Her favorite flavor was strawberry.

4. The ice-cream cone was invented in July at the 1904 St. Louis World's Fair.

ANSWERS

1. True. While it may seem boring to some, vanilla is number one, according to the International Ice Cream Association, with chocolate a distant second. (Over at Ben & Jerry's, Cherry Garcia was king in the summer of 2006, having overtaken the former leader, Chocolate Chip Cookie Dough.)

2. False. While Dolley served ice cream at the ball, it was already an established American treat from colonial days.

3. False. The actual inventor of this concoction is anonymous in history, but the name *sundae* is credited to Wisconsin ice-cream parlor owner George Giffy in the 1890s who thought it was a treat so special it should only be sold on Sundays. No one knows why the spelling of Sunday was changed.

4. Probably not. While credit for the cone has always gone to several ice-cream concessionaires in St. Louis in the summer of 1904, a patent for a cone-making device was issued in 1903 to New Yorker Italo Marchiony, a lemon ice seller, according to ice-cream history revisionists.

DON'T KNOW MUCH ABOUT
Hot Dogs

SOME SAY "HOT DOG." Some prefer "frankfurter." A few of us even say "weiner." In Coney Island, where they once banned the term "hot dog," you still order "foot longs." But whatever you call them, they are a summertime staple at backyard cookouts and baseball games. Americans eat about 7 billion "dogs" between Memorial Day and Labor Day. But how American is this classic picnic delight? Relish (Hold the mustard, please!) this quick quiz.

TRUE OR FALSE?

1. The frankfurter was named after Chicago's Frank Meat Company, which first produced them.

2. The name "hot dog" is derived from the fact that the earliest version of these sausages was made from man's best friend.

3. The world's longest hot dog reached 1,996 feet.

4. No legal "hot dog," Felix Frankfurter was the only naturalized American ever to serve on the Supreme Court.

ANSWERS

1. False. The frankfurter is named for Frankfurt, Germany, where it was first made more than five hundred years ago, although Vienna (Wien) also stakes a claim. It arrived in America around 1880, brought to St. Louis by Antoine Feuchtwanger, an immigrant from Frankfurt (which itself means "ford of the Franks").

2. False. Sports cartoonist Tad Dorgan sampled a "red hot" at New York's Polo Grounds and drew a cartoon featuring one of these "hot dogs."

3. True. It was made in honor of the 1996 Olympic Games in Atlanta.

4. True. Associate Justice Felix Frankfurter was born in Vienna. He was appointed to the Supreme Court in 1939 where he served until his retirement in 1962.

DON'T KNOW MUCH ABOUT
Popcorn

THINKING OF OCTOBER conjures up foliage, tailgating, beer fests, and Halloween. Well, get out your poppers, because October is also Popcorn Poppin' Month. (Some purists actually celebrate the last week in October as National Popcorn Week.) Long before butter topping and outsized tubs of popcorn at the multiplex, there was popcorn. Grown in China and other parts of Asia since ancient times, popcorn probably originated in Mexico. Ears of popcorn discovered in a New Mexico cave are thought to be more than 5,600 years old. Today, Americans consume more than 17 billion quarts of popped popcorn annually—about sixty-eight quarts for each of us. Hold the butter? What else do you know about this American favorite? Try this "pop" quiz.

1. What puts the "pop" in popcorn?

2. Does all corn "pop"?

3. Who introduced Europeans to popcorn?

4. Popping corn—vegetable or grain?

5. Is microwave popcorn just like regular popcorn?

ANSWERS

1. A drop of water surrounded by starch. As the water in a kernel of corn is heated, the water turns to steam and the pressure builds until it explodes and the soft starch inside inflates, turning the kernel of corn inside out.

2. Nope. There are six different types of corn, including popcorn: sweet, dent, flour, flint, and pod, but only popping corn pops.

3. The natives tried to sell popcorn to Columbus when he arrived and in 1519, Cortés got his first taste from the Aztecs. History shows the Native Americans of New England introduced popcorn to the Pilgrims in Massachusetts. There is some debate over the date of this historic event. Some say the Wampanoags brought a gift of popcorn to the first Thanksgiving; others say the recipe was turned over on February 22, 1630.

4. Popcorn is a cereal grain.

5. Yes and no. According to the Popcorn Board, microwave popcorn is the same as other popcorn except the kernels are usually larger.

DON'T KNOW MUCH ABOUT
Peanut Butter

DO YOU LIKE SMOOTH or crunchy? Are you a classic grape jelly purist? Or do you favor a more exotic preserve? Whatever your preference, chances are you like peanut butter. Half of the 2.4 billion pounds of peanuts consumed in America yearly are eaten as peanut butter. While we generally think of November as Turkey and Cranberry Month, this is actually National Peanut Butter Lovers' Month. (Not to be confused with National Peanut Month, which is March.) If your tongue's not stuck to the roof of your mouth, try this quick quiz.

1. The peanut: nut or pea? Which is it?

2. What famous American is considered the father of the peanut industry?

3. What state grows the most peanuts?

4. Are peanuts nutritious?

5. What style is more popular: smooth or chunky?

ANSWERS

1. The peanut is a legume—it bears fruit in the form of pods that contain one or more seeds. So the peanut is more closely related to peas than to nuts.

2. No, not Mr. Peanut! Born a slave, George Washington Carver (1864?–1943) was an American scientist famed for his agricultural research, especially his work with peanuts. Carver made more than three hundred products from peanuts, including a milk substitute, face powder, printer's ink, and soap.

3. Georgia produces more peanuts than any other U.S. state, about 40 percent of the country's annual peanut crop. Africa and Asia grow about 90 percent of the world's peanuts and the other leading peanut-growing countries are China, India, Indonesia, and Nigeria.

4. There are more energy-giving calories in roasted peanuts or peanut butter than in an equal weight of beefsteak.

5. According to industry sources, consumers prefer creamy by a 60–40 ratio. Children and women prefer creamy while men go for the crunchy stuff.

DON'T KNOW MUCH ABOUT
Rice

SEPTEMBER IS National Rice Month, a celebration of one of the world's most important food crops. Farmers grow rice in more than one hundred countries, and more than half of the world's population eats this grain as the main part of their meals. In some languages, the word *to eat* is the same as the word *to eat rice*. Short or long, brown or white, rice does more than "snap, crackle, and pop." What else do you know about this grain that helps feed the world?

1. Where did rice originate?

2. What country grows the most rice?

3. How long has rice been in America?

4. Why do people throw it at weddings? (And is that bad for birds?)

5. What is wild rice?

6. Who drinks their rice?

7. What classic American rice product was introduced in 1928?

Answers

1. No one knows exactly, but it probably first grew wild and was gathered in southeast Asia thousands of years ago. People cultivated rice for food by about 5000 B.C. in southern China and the northern part of Thailand, Laos, and Vietnam. In several languages in this area, the general terms for rice and food, or for rice and agriculture, are the same.

2. Not surprisingly, the world's most populous country, China, which grows 35 percent of the world's rice. India is the world's second leading rice producer and Asian farmers grow about 90 percent of the world's rice. The United States produces about 1 percent of the world's rice.

3. Rice was introduced to the New World by Europeans. The Portuguese carried it to Brazil, and the Spanish introduced it to several locations in Central and South America. The first record for North America dates from 1685, when the crop was produced in what is now South Carolina, carried to the area by slaves brought from Madagascar.

4. Throwing things at a newly married couple for good luck dates back to ancient Rome or Egypt. Most of the items thrown, including rice, are fertility symbols. There is no truth to the urban legend that uncooked rice eaten by birds will cause them to explode. Uncooked rice doesn't harm birds who often eat it in the wild.

5. In spite of its name, this grain is a type of grass that grows in central Canada and parts of the northern United States and is not closely related to rice.

6. In Japan, rice kernels are used to make a wine called sake. The Chinese also make rice wine. Rice is also used to make a nondairy alternative to milk called rice milk.

7. Kellogg's Rice Krispies was introduced in 1928.

DON'T KNOW MUCH ABOUT

DON'T KNOW MUCH ABOUT
Tea

DRINK UP! It's been over one hundred years since the invention of iced tea at the 1904 St. Louis World's Fair. British tea promoter Richard Blechynden wasn't selling much product in the heat. He plopped in some ice and invented a new summer favorite. Of course, tea is both ancient and popular. What do you know about the drink that helped start a revolution when Britain placed a tax on tea that led to the Boston Tea Party in 1773?

TRUE OR FALSE?

1. People have been drinking tea for nearly five thousand years.

2. There are more than twenty kinds of non-herbal tea.

3. After coffee and soft drinks, tea is the world's third most popular beverage.

4. Iced tea is more popular than hot tea in America.

5. The teabag is also more than one hundred years old.

6. Tea has more caffeine than coffee.

7. The British consume the most tea per capita.

8. China is the world's leading tea producer.

ANSWERS

1. True. According to legend, it was invented by a Chinese emperor in 2737 B.C.

2. False. There are just three: Green, Oolong, and Black tea. There is a very rare tea which some experts consider to be a fourth type of tea—White tea. In the United States about 94 percent of the tea consumed is Black.

3. False. Next to water, tea is the world's most popular beverage. In the U.S., it trails water, soft drinks, coffee, beer, and milk.

4. True. Over 80 percent of the tea consumed within the United States is the iced version.

5. True. In 1904, tea and coffee merchant Thomas Sullivan was credited with the accidental discovery of teabags. He sent out tea samples in small silk sacks. Orders soon began to pour in for tea in little bags.

6. True and false. On a dry weight basis, tea has twice as much caffeine per pound as does coffee, but brewing changes that. According to FDA figures, a 5-ounce cup of tea contains less caffeine than a 5-ounce cup of coffee.

7. False. The Irish consume more tea on a per capita basis.

8. False. India is the largest producer of tea, followed by China, Sri Lanka, and Japan.

DON'T KNOW MUCH ABOUT
Eggs

WHETHER THE CHICKEN came first or not, May is National Egg Month. No matter how you like yours, recent research has put eggs back on the American menu. The American Heart Association now says that for healthy individuals, an egg a day is fine. Still wondering which came first? Scramble or fry this quick quiz.

TRUE OR FALSE?

1. Egg color is based on the color of the hen.

2. A hen typically produces an egg every three days.

3. Christopher Columbus brought the chicken to the Americas.

4. On the vernal equinox, an egg will stand upright on its end.

5. Raw eggs wobble instead of spin.

6. Eggs Benedict are named for the famous traitor Benedict Arnold.

ANSWERS

1. True. White shelled eggs are produced by hens with white feathers and earlobes. Brown shelled eggs are produced by hens with red feathers and red earlobes. Yolk color depends on diet.

2. False. A hen requires twenty-four to twenty-six hours to produce an egg

3. True. Although there may have been chickens in America long ago, today's egg layers are related to chickens carried on Columbus's second voyage.

4. False, although widely believed. While some people have reported success with this experiment, it may be due to a peculiarity of the egg.

5. True. A test for whether an egg is hard-boiled is to spin it. If the egg spins easily, it is hard-cooked, but if it wobbles, it is raw.

6. False. According to legend, the dish was devised as a hangover remedy for a guest at the Waldorf-Astoria hotel and later named in his honor.

DON'T KNOW MUCH ABOUT
Garlic

SMELL THAT? It must be California's Gilroy Garlic Festival, which attracts more than 125,000 *Allium sativum* lovers and is held the last weekend in July each year in the self-proclaimed "Garlic Capital of the World." Of course, not everyone loves garlic, though it is hard to understand why. This wonderful herb is a staple of many international cuisines. Mentioned in the Bible, it fed the workmen who built the pyramids and was considered a powerful pre-battle food by Roman armies. So how did garlic get such a bad reputation? Nose around this quick quiz.

1. What is garlic?

2. What is a popular nickname for garlic?

3. What is the popular natural remedy for so-called "garlic breath"?

4. Will garlic really keep Dracula away?

Answers

1. It is a bulbous plant of the genus *Allium*, which includes onions, leeks, and shallots.

2. "Stinking rose" because of its strong odor.

3. Eating parsley, which is included in many garlic recipes.

4. We can't be sure. But garlic's reputation as a vampire deterrent goes way back, long before Bram Stoker wrote about it in his classic *Dracula*. The reputation of garlic as a vampire deterrent goes back in history. It may be related to the idea that garlic is a mosquito repellent, another notorious bloodsucker.

DON'T KNOW MUCH ABOUT
Cherries

IN CASE YOU hadn't heard, February is National Cherry month. For some reason, the cherry hasn't acquired the same traditional respectability that apples have, even though the cherry arrived in North America with some of the first European settlers. French colonists from Normandy are credited with planting the first cherry trees along the St. Lawrence River and near the Great Lakes. While the old saying about "an apple a day" is well known, recent research suggest that cherries are also beneficial, possessing valuable antioxidants. Think life is just a bowl of cherries? Pick on this quick quiz.

1. Did George Washington really chop down his father's cherry tree?

2. Cherries may have played a role in the death of which president?

3. What Michigan location is the Cherry Capital of the World, host to the National Cherry Festival in July?

4. Who said, "Violence is as American as cherry pie"?

Answers

1. No. This is a legendary tale concocted by Mason Weems who wrote a highly embellished biography of George Washington in 1800, filled with moralizing stories for children. The cherry tree episode was one of his inventions.

2. Zachary Taylor, the twelfth president, who died on July 9, 1850. After supposedly wolfing down a bowl of cherries and a pitcher of ice milk, Taylor became violently ill. The cause of death was cholera. Perhaps poor sanitation in the nation's capital made raw fruit and fresh dairy a risky meal in ninteenth-century Washington.

3. Traverse City, Michigan. Michigan leads the nation in cherry production, producing about three-quarters of the U.S. crop.

4. Black radical leader H. Rap Brown in July 1967.

CIVICS

DON'T KNOW MUCH ABOUT
Constitutional Amendments

"No law, varying the compensation for the services of the Senators and Representatives, shall take effect, until an election of Representatives shall have intervened." In other words, Congress can't vote itself a raise. In 1992, this became the law of the land as the 27th Amendment, the most recent constitutional amendment. But it took awhile for the idea to wind its way through the system—about two centuries! What else do you know about the amendments, those rare changes to America's Constitution? You should know. After all, these rights belong to you. Ratify this quick civics refresher.

1. Who wrote the 27th amendment?

2. What are the first ten amendments called?

3. How many amendments have been proposed by presidents?

4. How does a constitutional amendment get ratified?

5. Which constitutional amendment cancels out an earlier amendment?

ANSWERS

1. James Madison, one of the architects of the Constitution and Bill of Rights. It was actually one of the first amendments ever proposed. The states that had originally voted to ratify this amendment in the late 1700s were still counted and did not have to vote to ratify again two hundred years later.

2. The first ten amendments are called the Bill of Rights. Many states would not agree to ratify the Constitution unless guaranteed that a Bill of Rights was added to the Constitution.

3. None. While the president can certainly change opinion about an amendment, the power to amend is legislative.

4. A two-thirds majority in both the House and Senate are required to submit an amendment to the states for ratification. Three-quarters of the states must then approve the proposal for ratification. Although state legislatures usually decide on ratification, a state convention can also be called.

5. In 1933, Amendment 21 repealed Amendment 18, which prohibited the manufacture, sale, and transportation of liquor in the U.S. in 1919.

DON'T KNOW MUCH ABOUT
the New Hampshire Primary

WITH NEW HAMPSHIRE'S traditional first-in-the-nation primary election, the tiny Granite State (population 1.3 million in 2005) holds center stage in the political spotlight. Although the primary tradition in New Hampshire dates back to 1913, the impact of this primary has grown in modern times, and few presidents have been elected without tramping through the state's wintry landscape. What do you know about this American political tradition? Cast a vote in this quick quiz.

1. How old is the New Hampshire primary and was it always first?

2. How many men have won the presidency without winning their party's New Hampshire primary?

3. Who has won the most New Hampshire primaries?

4. Which New Hampshire victor surprisingly dropped out of the presidential race?

5. Can Independent voters vote in New Hampshire's primary?

6. How many natives of New Hampshire became president?

7. What was New Hampshire's distinction in the 2000 presidential general election?

ANSWERS

1. The first recorded primary was in 1913, but the modern tradition dates to 1952. New Hampshire has been first since then, and state law requires that its primary remain first in the nation.

2. Two so far. In 1992, Bill Clinton finished second to Senator Paul Tsongas of neighboring Massachusetts and went on to win the presidency; in 2000, George W. Bush lost New Hampshire to Senator John McCain of Arizona but still won the nomination and the presidency.

3. Richard Nixon, in 1960, 1968, and 1972.

4. In 1968, President Lyndon B. Johnson did not file for the primary, but won anyway with write-in votes. But Senator Eugene McCarthy, who campaigned actively against the Vietnam war, won 41 percent of the vote and gained more delegates than the president. Johnson was so stunned that he soon announced that he would not run for reelection.

5. Republicans and Democrats are restricted to voting only for the candidates of their own party, but voters who choose to register as Independents can vote in either the Republican or Democratic primary.

6. Only one, Franklin Pierce. The fourteenth president, he served from 1853 to 1857.

7. New Hampshire, with four electoral votes, was the only New England state carried by George Bush. Had Al Gore won the state, he would have won the election.

DON'T KNOW MUCH ABOUT
the Supreme Court

THINK THAT THE Supremes made "Baby Love" and a lot of other great Motown hits? They did, but the other Supremes take center stage on the "First Monday in October," the traditional beginning of a new Supreme Court session. From abortion to elections, the death penalty to eminent domain, the Supreme Court is at the center of the most significant issues facing America. Congress and presidents come and go, but the impact of its decisions can last for decades or centuries. Yet these nine people in black robes and what they do is a mystery to many Americans. What do you know about the history, traditions—and mythology—of the highest court in the land?

1. Did the Constitution originally specify nine justices?

2. Does the Constitution say that the Supreme Court decides which laws are unconstitutional?

3. How often do nominees to the court not make it?

4. What is the "Rule of Four"?

5. The Chief Justice presides over a president's impeachment trial. But has a Supreme Court justice ever been impeached?

6. Who was the longest serving justice in history? The longest serving chief justice?

7. Who is the only president since Franklin D. Roosevelt not to appoint a Supreme Court justice?

ANSWERS

1. No, the original Court had six—five associates and a chief justice. Congress changed the number of justices seven times until settling on nine in 1869. It has been that way ever since.

2. Not specifically. This power, known as "judicial review," was established in 1803 in a case known as *Marbury v. Madison*, and has become a basic part of the American legal system.

3. When White House Counsel Harriet Meiers withdrew her nomination to the Supreme Court in October 2005, she became the seventh person to withdraw. A total of thirty nominees have failed to make it onto the court, either by withdrawing or being rejected by the Senate. Most of those failures came before the twentieth century, and eight of President John Tyler's nominees were unsuccessful, a record that may go unmatched.

4. The unwritten tradition that a case will not be heard by the Supreme Court unless at least four justices vote to review it.

5. Yes, in 1804, Samuel Chase was impeached—equivalent to being indicted—but not convicted. One other justice, Abe Fortas, resigned when threatened with impeachment in 1969.

6. William O. Douglas served for thirty-six years; John Marshall served as chief justice for thirty-four years.

7. Jimmy Carter.

IT WAS IN DECEMBER 2000, that the Supreme Court made
the historic ruling that made George W. Bush president.
Shrouded in more secrecy than any other branch of government, the Supreme Court can be a mysterious and confus-
ing place. What else do you know about the highest court in
the land? "Hear Ye, Hear Ye," take this bonus quiz.

1. Who was the first chief justice?

2. Who was the first black justice?

3. Who is the first woman justice?

4. What former president became chief justice?

5. The chief justice need not be an associate. How many
   chief justices were associate justices?

ANSWERS

1. John Jay.

2. Thurgood Marshall, appointed in 1967.

3. Sandra Day O'Connor, appointed in 1983.

4. William H. Taft.

5. Only three: Edward White, Harlan Stone, and William Rehnquist.

DON'T KNOW MUCH ABOUT
Presidential Elections

If twentieth-century history is a guide, being vice president is the way to go to become president. Of the eighteen presidents in the twentieth century, six were sitting "veeps" who made the move to the White House. The bad news? Only one actually won the office on his own. The other best path to the Oval Office? Being governor. Six twentieth-century presidents were governors before their election. What do you know about becoming president?

1. Which twentieth-century sitting vice presidents became president?

2. Name the six who went from Governors' mansions to the White House.

3. Since George Washington, leading troops was considered a stepping-stone for many presidents. How many twentieth-century presidents were generals? And how many have served in the military?

4. What do Jack Kemp, Richard Gephardt, Phil Crane, and Morris Udall have in common?

ANSWERS

1. Teddy Roosevelt, Coolidge, Truman, and Johnson by death of the president; Ford by resignation. George Bush was vice president when elected in 1988. Richard Nixon was defeated in 1960 as the sitting vice president, but made it in 1968 on his second try.

2. McKinley, Wilson, Franklin D. Roosevelt, Carter, Reagan, and Clinton. Texas Governor George W. Bush continued the trend into the new century with his election in 2000.

3. One. Eisenhower. Ten others with some military credentials were McKinley, Teddy Roosevelt, Truman, Kennedy, Johnson, Nixon, Ford, Carter, Reagan, George H. W. Bush, and George W. Bush (Air National Guard).

4. All ran unsuccessfully for president from the House of Representatives. No House member was elected president in the twentieth century.

DON'T KNOW MUCH ABOUT
Presidential Inaugurations

Just before the first inauguration, George Washington suggested to a friend that the president be called "His High Mightiness." Congress decided otherwise. In that first ceremony, held in New York in April 1789, Washington did improvise a bit, adding "So help me God" to the constitutional oath. Every president since has honored that tradition. What do you know about some of America's most notable inaugurals? Raise your right hand for this quick quiz.

1. Which president had to escape the inaugural celebration when the crowd got too rowdy?

2. The longest inaugural address ever was also deadly. Which president died a month after taking the oath?

3. Which president secretly arrived in Washington for his inauguration under an assumed name?

4. Who said, "The great government we loved has too often been made use of for private and selfish purposes"?

5. Who said, "The only thing we have to fear is fear itself"?

6. Who said, "Ask not what your country can do for you — ask what you can do for your country"?

ANSWERS

1. Andrew Jackson in 1824. When hundreds of westerners in buckskin clothes and muddy boots mobbed the White House, the new president fled out the back door. Luckily, someone thought to move the punch buckets out to the lawn.

2. William Henry Harrison delivered a one hour and forty-minute address in the cold on March 4, 1841. He did not wear a hat, gloves, or overcoat and caught a severe cold. A month later on April 4, Harrison died of pneumonia, the first president to die in office.

3. Abraham Lincoln. With tensions high, the danger was already great when Lincoln came to Washington in 1860. He arrived in secret to avoid an assassination plot.

4. Woodrow Wilson, first inaugural, March 1913.

5. Franklin D. Roosevelt, first inaugural, March 4, 1933.

6. John F. Kennedy, inaugural, January 20, 1961.

PRESIDENT BUSH took the oath of office for a second term on January 20, 2005, as required by the Constitution's 20th Amendment. It wasn't always that way. A blend of traditions and law, the presidential inaugural has been a movable feast since the first one 216 years ago, when George Washington established quite a few presidential precedents. What else do you know about this banner day on the American calendar? "Swear to or affirm" your knowledge in this bonus quiz.

1. Who took the first outdoor oath of office?

2. Washington did two things not required by law that most other presidents have also done. What were they?

3. Who was the first inaugurated at the Capitol Building in Washington, D.C.?

4. Who made the longest inaugural address? The shortest?

5. Which president was first to neither swear the oath nor *kiss* the Bible?

6. In which inaugural parade did African Americans first participate? The first parade with women?

Answers

1. Washington at the first inaugural in 1789 in New York City, then briefly the nation's capital.

2. He said "So help me God" and then kissed the Bible after taking the oath.

3. Thomas Jefferson in March 1801.

4. William Henry Harrison with ten thousand words on March 4, 1841; it was followed by the shortest presidency. Harrison died of pneumonia on April 4, the first president to die in office. Washington gave the shortest address; his second was a mere 135 words.

5. Franklin Pierce (1853) who "affirmed" rather than swore the oath as the Constitution allows. He merely placed his left hand on the Bible.

6. African Americans marched in Lincoln's second inaugural (1865); women waited until Woodrow Wilson's second inaugural in 1917.

DON'T KNOW MUCH ABOUT
the Pentagon

THE SEPTEMBER 11, 2001 attacks struck deadly blows at two of America's most familiar landmarks. One of them, the Pentagon, was completed on January 15, 1943. Headquarters of the Department of Defense, the Pentagon is one of the world's largest office buildings. Lying on the west bank of the Potomac River in Arlington, Virginia, across from Washington, D.C., it was built in the form of a five-sided figure to house the scattered offices of what was then called the War Department under one roof. What else do you know about this architectural wonder? Enlist in this quick quiz.

TRUE OR FALSE?

1. More than one hundred thousand people work in the Pentagon each day, only a few of them civilians.

2. Begun before America entered World War II, the Pentagon was completed in less than a year and a half.

3. With five floors, a mezzanine, and a basement, the Pentagon has nearly 4 million square feet of office space.

4. At a protest during the Vietnam War, chanting antiwar demonstrators attempted to surround the Pentagon and make it "levitate."

5. *The Pentagon Papers* is a daily newspaper published by the Defense Department for its employees.

ANSWERS

1. False. According to the Pentagon in 2006, approximately twenty three thousand employees and another three thousand non-defense personnel work in the Pentagon each day; about half of these are civilians.

2. True. Army engineers began work in September 1941, and completed the job in sixteen months at a cost of $83 million.

3. True. The building covers twenty-nine acres and has about 3,706,000 square feet of office and other space. It stretches about one mile around. Parking areas adjacent to it cover sixty-seven acres and can hold about ten thousand vehicles. Restaurants and cafeterias in the Pentagon serve over fifteen thousand meals daily. The building also has shops, a radio and television station, bank, dispensary, post office, and heliport.

4. True. According to Norman Mailer's *Armies of the Night.* "In the air, the Pentagon would then, went the presumption, turn orange and vibrate. . . . At that point, the war in Vietnam would end."

5. False. This is the popular name given to a set of secret documents regarding America's involvement in Vietnam. When they were leaked to newspapers, the U.S. government attempted to stop publication, but the Supreme Court ruled in favor of the press in a landmark First Amendment decision.

DON'T KNOW MUCH ABOUT
the FBI

THERE ARE SOME exclusive lists you don't want your name on. On March 14, 1950 the FBI created its famed Ten Most Wanted list after a reporter requested a list of the "toughest guys" wanted by the Federal Bureau of Investigation. Thomas Holden was the first man to make the list. A bank robber and murderer, he was captured in 1951 and spent the rest of his life in prison. Born in 1908 as the Bureau of Investigation, the FBI was given its current name seventy years ago by Congress. What else do you know about the bureau and its notorious list?

1. Who served as director from 1924 to 1972?

2. Who appoints the director?

3. Who has the current highest reward offered on the Most Wanted list?

4. True or False: Fewer than ten women have made the Most Wanted list.

5. True or False: A college degree is required to become an FBI Agent.

6. True or False: There is a Most Wanted list for stolen art.

ANSWERS

1. J. Edgar Hoover, who served until his death in 1972. In 1975, a Senate committee reported that under Hoover, FBI agents had committed burglaries and spied illegally on U.S. citizens. Senate investigators also charged that Hoover had given certain presidents damaging personal information about their political opponents.

2. The president nominates the director who must be confirmed by the Senate. In 1976, Congress limited the term of the FBI director to ten years.

3. Osama Bin Laden, at more than $25 million.

4. True. Seven women have appeared on the list over the years.

5. True. A four-year degree is a prerequisite.

6. Sort of. The FBI maintains a National Stolen Art File. In the summer of 2006 it included *The Scream* and another piece by Edvard Munch since recovered, looted and stolen Iraqi artifacts, and a $3 million Stradivarius violin stolen in New York in 1995 from a noted concert violinist.

WHAT ELSE DO YOU KNOW about the bureau and its notorious list? Take this bonus quiz.

1. How old is the FBI?

2. Who was in charge when the Ten Most Wanted list first appeared in 1950?

3. Who was the director of the FBI on September 11, 2001 during the attacks?

4. What notorious foreigner is currently on the list?

ANSWERS

1. It originated in 1908 as the Bureau of Investigation, was reorganized in 1924, and acquired its present name in 1935.

2. J. Edgar Hoover, the first director of the FBI, who knew a good publicity angle when he saw one.

3. Robert S. Mueller, III, became the sixth director of the FBI on September 4, 2001, just one week before the terrorist attacks.

4. Suspected terrorist mastermind Osama Bin Laden.

DON'T KNOW MUCH ABOUT
the Democratic Party

TIMES HAVE CHANGED since political nominating conventions were long, dramatic affairs that actually produced a nominee, sometimes after many ballots and deals were cut in "smoke-filled rooms." The primary system has reduced political conventions to pageants that usually confirm the obvious. What do you know about America's oldest political party? Register for this quick quiz.

TRUE OR FALSE?

1. Thomas Jefferson founded the Democratic Party in 1793.

2. The first Democratic nominating convention took place in 1832.

3. The donkey is the party's official emblem.

4. After the violence at the controversial 1968 Chicago Democratic convention, an official report called the incident "a police riot."

5. The twenty-year stretch of Democrats from Roosevelt to Truman (1933–1953) is the longest period that one party has held the White House.

ANSWERS

1. False. While the party has roots in Jefferson's Democratic-Republican Party, historians link the party's beginnings to Andrew Jackson who defeated Federalist John Quincy Adams in 1828.

2. True. They chose incumbent Andrew Jackson. It was not the first convention. The Anti-Masonic Party, a third party, held the first nominating convention in 1831.

3. False. Unlike the Republicans, who have used the elephant as their official party symbol, the donkey has never been officially adopted by the Democrats. The widely used emblem first appeared in an 1870 political cartoon by Thomas Nast.

4. True. The Walker Report, part of a commission investigating urban rioting, used that term to describe the pitched battles between antiwar protesters and Chicago police.

5. True and False, depending on how you count Andrew Johnson. Republicans held the presidency for twenty-four years from Lincoln (1861) through Chester Arthur (1865). Johnson, the vice president who replaced Lincoln after his death, was a Democrat, but he was nominated and elected on the Republican ticket.

DON'T KNOW MUCH ABOUT
the Electoral College

If thinking about the Electoral College gives you the chills, you are not alone. Grown men tremble when asked what the Electoral College is and who is in it. For years it was an abstraction until the election of 2000, when we discovered that it does count for something in American politics. Think the Electoral College is a "party school"?

1. How many electors are there?

2. What determines the number of electors?

3. What is the fewest number of electors a state can have?

4. What happens if there is a tie or no one wins enough electors?

5. Before 2000, how many times had the popular winner lost the presidency?

6. Who was the last third party candidate to win any electors?

ANSWERS

1. 538.

2. There is one elector for every member of Congress—435 in the House, 100 in the Senate—and three electors for the District of Columbia. Two hundred seventy electoral votes are needed to win.

3. Three, because every state has at least one Representative and two Senators.

4. The election goes to the House of Representatives where each state delegation would get a single vote.

5. Three times: 1828, 1876, and 1888.

6. George Wallace in 1968.

DON'T KNOW MUCH ABOUT
the Republican Party

ON FEBRUARY 28, 1854, fifty-three men met in Ripon, Wisconsin, to begin organizing a new political party. Mostly northerners opposed to slavery moving into new U.S. territories, they decided to organize the Republican Party if Congress passed the Kansas-Nebraska Bill. That law, allowing new states to choose whether to be free or slave, was passed and the delegates formally adopted the name Republican in July 1854. (Claims of a first Republican Party are also made by New Hampshire, which had earlier formed a statewide Republican Party.) What else do you know about the party of Lincoln, Teddy, Reagan, and the Bushes? Register for this quick quiz.

TRUE OR FALSE?

1. GOP means "Get Out and Push."

2. The elephant is the party symbol because it "works for peanuts."

3. Lincoln was the first Republican president.

4. The GOP is the oldest party.

5. Since 1952, there have been more Republican presidents than Democrats.

ANSWERS

1. False. Shorthand for Republican, the term GOP dates back to the 1870s, but its meaning has changed over the years. In 1875, the *Congressional Record* referred to "this gallant old party," but there is another reference in 1876 to "Grand Old Party." "Get Out and Push" was used as a party slogan in the 1920s.

2. False. Famed political cartoonist Thomas Nast drew a cartoon which appeared in November 1874, in which a Democratic donkey clothed in a lion's skin scared away other animals, among them an elephant, labeled "The Republican Vote." Nast chose the elephant because it was thought to be steadfast and easily controlled, but unmanageable when frightened. Other cartoonists soon adopted the elephant to symbolize Republicans and Nast's donkey became the symbol of the Democrats.

3. True. In 1860, Abraham Lincoln became the party's first president, winning about 40 percent of the vote with the rest being split between three other candidates. The first Republican presidential candidate was General John C. Frémont, a dashing explorer and soldier, who lost in 1856.

4. False. Whether Gallant or Grand, it is also not the oldest, since the Democratic Party was organized in 1830 under Andrew Jackson and comes from the even earlier Democratic-Republicans of Thomas Jefferson's day.

5. True. Six Republicans (Eisenhower, Nixon, Ford, Reagan, and both Bushes) to four Democrats (Kennedy, Johnson, Carter, and Clinton). From 1952 through 2000, a Republican has won eight of the thirteen presidential elections. From 1860, when Lincoln was elected, to 1932, Republicans won fourteen of the eighteen presidential elections.

DON'T KNOW MUCH ABOUT
Presidential Campaign Slogans

PRESIDENTIAL CAMPAIGNS have altered the American vocabulary. Everything from "log cabins" to "booze" and "OK" got their start in presidential campaigns. ("OK" stood for "Old Kinderhook." Martin van Buren hailed from that small town in upstate New York.) Along the way there have been many memorable words and phrases added to the American language. "I Like Ike" may have been the perfect campaign slogan. It was short and sweet. A positive message. But presidential campaign slogans aren't always so nice—or memorable. Match these famous—or notorious—presidential campaign slogans with the candidates.

1. "He Kept Us Out of War."

2. "Rum, Romanism and Ruin."

3. "Back to Normalcy."

4. "A chicken in every pot and two cars in every garage."

5. "In Your Heart You Know He's Right."

ANSWERS

1. President Woodrow Wilson, reelection campaign of 1916. The U.S. would enter the war in 1917.

2. A Republican slogan aimed at Al Smith, the Roman Catholic governor of New York in his 1928 race with Herbert Hoover.

3. Warren G. Harding in 1920.

4. Herbert Hoover, 1932.

5. Barry Goldwater, the Conservative Republican who lost in a landslide to Lyndon Johnson in 1964.

DON'T KNOW MUCH ABOUT
American Mottoes

What do these famous American phrases mean and where do they come from?

1. "E Pluribus Unum."

2. "In God We Trust."

3. "Life, Liberty and the Pursuit of Happiness."

4. "We, the People . . ."

5. "Mind Your Business."

ANSWERS

1. This Latin phrase appears on the Great Seal of the United States, and means "From many, one." Ironically, it may have come from the British *Gentleman's Magazine*, widely popular in America during the 1700s. The motto was adopted by Congress in 1781.

2. This phrase was first used on American coins in 1864, during the Civil War. It was dropped and then later restored by Congress.

3. Three of the "certain unalienable Rights" specified in Jefferson's Declaration of Independence.

4. The opening words of the preamble to the U.S. Constitution, adopted in 1789.

5. The first motto that appeared on U.S. coins.

DON'T KNOW MUCH ABOUT
the Medal of Honor

FOR GENERATIONS OF Americans, October 8 was known as Alvin York Day in honor of World War I hero Alvin Cullum York (1887–1964). Born in Tennessee, York was an amazing marksman, and on October 8, 1918, York led seven men against a German machine gun nest. According to the official citation, York and his men attacked the Germans, killing 32 and then capturing more than 132 enemy soldiers. For his bravery York was awarded America's highest military decoration, the Medal of Honor (usually referred to as the Congressional Medal). What do you know about the nation's prized award, presented 3,460 times to date, "for conspicuous gallantry and intrepidity at the risk of life, above and beyond the call of duty, in actual combat against an armed enemy force."

TRUE OR FALSE?

1. When drafted, Alvin York applied for conscientious objector status.

2. The Medal of Honor was created during the Civil War.

3. No woman has even been awarded the Medal of Honor.

4. The first African American to receive the award came during Vietnam.

5. Medals of Honor have been presented in every major conflict since the Civil War.

ANSWERS

1. True. Deeply religious, York sought an exemption because he believed war was wrong, but was denied because his particular denomination was not recognized as a church at the time. Famed as "Sergeant York," he was a corporal at the time of his feat. Gary Cooper portrayed him in a 1942 film and won the Academy Award for Best Actor.

2. True. During the Civil War, Congress authorized a navy medal of valor and later one for the army. Today this award, often referred to as the Congressional Medal of Honor, is the highest military decoration that the United States grants to members of its armed forces. Originally, the Army Medal of Honor was awarded only to noncommissioned officers and privates, but, beginning in 1863, this honor was also given to officers.

3. False. Dr. Mary Edwards Walker, a nurse turned battlefield surgeon and later captured as a Union spy, received the award in 1865. During a purge of the awards in 1917, her medal was revoked but restored posthumously in 1977.

4. False. William Carney, a former slave and member of the Massachusetts 54th Colored Infantry, was the first African American to earn the medal for his actions at Fort Wagner, South Carolina, the battle depicted in the film *Glory*.

5. False. There were no awards during the actions in Grenada, Panama, Lebanon, or Desert Storm. As of spring 2007, the most recent had been given posthumously on April 4, 2005, to Sergeant Paul Smith for actions in Iraq.

AFTERWORD

OK, HOTSHOT. HOW DID YOU DO?

Are you ready to throw down the gauntlet to Alex Trebek and grab that *Jeopardy!* buzzer? Ready to say "Final Answer" with complete authority and become a Millionaire? Have you finally shown your teenagers that you're not as dumb as they think? Maybe you simply realized somewhere while reading this book that Dr. Spock was right: You do know more than you think you do.

Or is it really time to go back and crack those books again?

Maybe this "take home test" has left you feeling a little of both. Satisfied in what you do know, but eager and curious to learn more. If that is what I have accomplished—and we have also had a pretty good time along the way—then my work here is done. But you may still have some work to do.

Curiosity has driven many human beings to ask questions like:

Why does that ball of fire rise in the sky each day?

Where do those lights come from at night?

What's on the other side of that mountain?

What will happen to this lobster if I plunge it in boiling water?

Or, what would the world look like if I rode on a beam of light? (Well, okay. Albert Einstein asked that one. The question honestly never occurred to me.)

Asking questions and finding the answers has been the heart of human civilization. It is there from the earliest days when we ask, "Why is the sky blue?" and "Are we there yet?"

I don't know about killing the cat, but we have made it this far as a race on curiosity—the quest for knowledge, wisdom, insight, understanding, and the perfect hamburger. A little knowledge—or learning—can be a dangerous thing. But looking for the answers is a great adventure. Enjoy the ride, hotshot.

INDEX OF SUBJECTS

FAMOUS PEOPLE

Benjamin Franklin, 7
Walt Whitman, 9
Rosa Parks, 11
Malcolm X, 13
Houdini, 15
Mother Teresa, 17
Michelangelo, 19
Young George Washington, 21
Dr. Spock, 23
Gandhi, 25
Henry Ford, 27
Ralph Waldo Emerson, 29
J.R.R. Tolkien, 31
Sitting Bull, 33
Queen Elizabeth II, 35
Anne Frank, 37
Helen Keller, 39
Abraham Lincoln, 41

Amelia Earhart, 43
Thomas A. Edison, 45
Albert Einstein, 47

EXCEPTIONAL PLACES

Panama Canal, 51
Everglades, 53
Rocky Mountains, 55
Vietnam, 57
London, 59
Hiroshima, 61
Great Smoky Mountains, 63
Berlin, 65
Mexico, 67
Washington, D.C., 69
Mardi Gras in New Orleans, 71
Memorials, 73

HISTORIC HAPPENINGS

Emancipation
  Proclamation, 77
Lewis and Clark
  Expedition, 79
Watergate, 81
Nuremberg Trials, 83
St. Louis World's Fair, 85
LBJ's "War on Poverty," 87
Burr-Hamilton Duel, 89
D-Day, 91
World War I, 95
Arab Oil Embargo, 97
Korean War, 99
Civil Rights Movement, 101
Salem Witch Trials, 103
Great Crash, 105
Civil War, 107

HOLIDAYS AND TRADITIONS

New Year's Eve, 111
Easter Customs, 113
Passover, 115
Kwanzaa, 117
American Christmas
  Traditions, 119
Christmas Traditions, 121
Presidents' Day, 123
Irish-American History, 125
Independence Day, 129
Chinese New Year, 131
Valentine's Day, 133

Memorial Day, 135
Santa Claus, 137

EVERYDAY OBJECTS AND REMARKABLE INVENTIONS

New York's Subway System, 141
U.S. Highway System, 143
Digital Photography, 145
McDonald's, 147
Barbecue, 149
Paper Money, 151
Phonograph, 153
Microwave Ovens, 155
Atomic Bomb, 157
Automobile, 159

SPACE AND THE NATURAL WORLD

Pluto, 163
Moon, 165
Universe, 167
Women in Space, 169
Mir Space Station, 171
Moon Landing, 173
Asteroids, 175
Volcanoes, 177
Vernal Equinox, 179
Earthquakes, 181

SPORTS

NASCAR, 185
NFL, 187
Babe Ruth, 189

Ancient Olympics, 191
New York Yankees, 193
World Series, 195
Sex and Sports, 197
Basketball, 199
Jackie Robinson, 201
Muhammad Ali, 203

ENTERTAINMENT

Alfred Hitchcock, 207
George Gershwin, 209
Lucille Ball, 211
Psycho, 213
Disneyland, 215
Tonight Show, 217
Motown, 219
Playboy, 221
60 Minutes, 223
The Catcher in the Rye, 225
Walt Disney, 227
Mickey Mouse, 229
The Beatles, 231

FOOD

Cornflakes, 235
Hamburgers, 237

Ice Cream, 239
Hot Dogs, 241
Popcorn, 243
Peanut Butter, 245
Rice, 247
Tea, 249
Eggs, 251
Garlic, 253
Cherries, 255

CIVICS

Constitutional Amendments, 259
New Hampshire Primary, 261
Supreme Court, 263
Presidential Elections, 267
Presidential Inaugurations, 269
Pentagon, 273
FBI, 275
Democratic Party, 279
Electoral College, 281
Republican Party, 283
Presidential Campaign Slogans, 285
American Mottoes, 287
Medal of Honor, 289

ACKNOWLEDGEMENTS

A great many friends and colleagues helped me pull this project together and I am very grateful to all of them. Among the many people who have been so supportive at HarperCollins, I would like to thank Leslie Cohen, Phil Friedman, Jen Hart, Hope Innelli, Carrie Kania, Andrea Rosen, Avery Schlicker, Michael Signorelli, Joe Tessitore, and freelance publicist Laura Renolds.

All of the people at the David Black Literary Agency are not only industrious, smart, and good-looking, they are also my very good friends, and my life and work have been made better by David Black, Leigh Ann Eliseo, Dave Larabell, Gary Morris, Susan Raihofer, and Joy Tutela.

This material started out years ago with a single quiz in *USA Weekend*, where it was my pleasure to work with Jack Curry and Tom Lent, who have since moved on. My gratitude to them and the rest of the staff at the magazine.

My family is always my greatest inspiration and so my thanks and love go out to Colin and Jenny. And to my wife Joann Davis who makes it all possible and worthwhile.

—Dorset, Vermont
May 2007

BOOKS BY KENNETH C. DAVIS

DON'T KNOW MUCH ABOUT MYTHOLOGY
Everything You Need to Know About the Greatest Stories
in Human History but Never Learned
ISBN 0-06-019460-X (hardcover)
ISBN 0-06-093257-0 (paperback)
Explores the great legends of human history and connects
them to historical events.

DON'T KNOW MUCH ABOUT HISTORY
Everything You Need to Know About American History but Never Learned
ISBN 0-06-008382-4 (paperback)
The million-copy *New York Times* bestseller, completely revised
and updated! With wit, candor, and fascinating facts, Davis
explodes long-held myths and misconceptions
of American history.

DON'T KNOW MUCH ABOUT THE BIBLE
Everything You Need to Know About the Good Book but Never Learned
ISBN 0-380-72839-7 (paperback)
Davis sets the panorama of the Scriptures against the historical
events that shaped them.

DON'T KNOW MUCH ABOUT THE CIVIL WAR
Everything You Need to Know About America's
Greatest Conflict but Never Learned
ISBN 0-380-71908-8 (paperback)
Davis sorts out the players, politics, and key
events of the Civil War.

DON'T KNOW MUCH ABOUT GEOGRAPHY
Everything You Need to Know About the World but Never Learned
ISBN 0-380-71379-9 (paperback)
A fascinating, breathtaking, and entertaining grand tour of the
planet Earth.

DON'T KNOW MUCH ABOUT THE UNIVERSE
Everything You Need to Know About Outer Space but Never Learned
ISBN 0-06-093256-2 (paperback)
Davis begins with 20th-century exploration and coverage of
celestial bodies, and concludes with an accessible exploration of
cosmic questions from the beginning of civilization to the present.

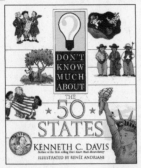
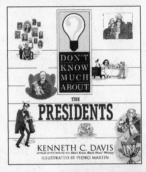

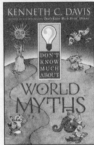